BIOLOGICAL WEAPONS

ENCYCLOPAEDIA OF BIOTERRORISM

Vol. III

BIOLOGICAL WEAPONS

By

Dr. S.K. Prasad

School of Studies of Zoology & Biotechnology

Vikram University

Ujjain

DISCOVERY PUBLISHING HOUSE PVT. LTD.

NEW DELHI-110 002

First Published-2008

ISBN 978-81-8356-302-4 (Set)

Published by:

DISCOVERY PUBLISHING HOUSE PVT. LTD.

4831/24, Ansari Road, Prahlad Street,
Darya Ganj, New Delhi-110002 (India)
Phone: 23279245 • Fax: 91-11-23253475
E-mail: dphbooks@rediffmail.com
dphtemp@indiatimes.com

Printed at:

Sachin Printers, Delhi

Preface

The present title *"Encyclopaedia of Bioterrorism"* aims to bring historical context to present concerns about biological weapons, biological agents, chemical weapons and the potential for bioterrorism. The lack of use of biological weapons in war advocates for using biology to create a new class of weapons initially envisioned delivery systems for pathogenic aerosols that mimicked those for chemical weapons, which were mainly bombs that generated aerosols intended to kill or disable troops in a local area. This vision was quickly replaced by the concept of creating huge clouds of germs that would drift with the wind and infect people over areas of thousands of square miles. The scientists and civil and military leaders who believed in the future of biological weapons saw their potential for fulfilling the goals of total war, for the mass killing or debilitation of enemy civilians. It is often asked why biological weapons are different from any other means of destruction. The answer is that they are the only ones devised expressly to kill defenseless humans and animal and plant life, with little real battlefield potential in modern war.

When biological weapons were developed for possible retaliation against an enemy thought to be similarly armed, they fit this model of restraint. Nevertheless, germ weapons were developed and laws and political circumstances offered no guarantees against their use. Historians may seek neat explanations, but the element of uncertainty was always there. Political and military authorities differed unpredictably when it came to calculating the consequences of using biological weapons. This book also describes the new subject of bioterrorism and traces at least the beginnings of the present era, in which domestic preparedness and homeland security are major policy issue. One of the major homeland security directives is to use technology derived from medical research to protect civilians against bioterrorism. The use of biology to defined civilians against any and all biological agents is a daunting project that imposes new security restrictions more familiar to physicists in the defense establishment than to biologists.

The study of biological weapons combines knowledge from disparate fields: biology, medicine, military history, politics, law, and ethics. This book intends to give the reader a basic literacy in this complex area. The entire subject of biological weapons is characterized by an unusual degree of misinformation and even disinformation. Almost every fact in this book has several page backstory, with greater nuance and depth than a brief overview can provide. As scholars continue their work, more information and analysis will likely turn today's accepted wisdom on its head. This progress will be a healthy sign for the field, whose subject matter has often been exploited for the frightening and sometimes entertaining effect it has on the imagination.

There can be no claim to originality except in the manner of treatment and much of the information has been obtained from the books and scientific journals available in the different libraries.

The author expresses his thanks to his friends and colleagues whose continue inspirations have initiated him to bring out this book.

The author expresses his gratitude to Mr. Wasan and staff of M/s Discovery Publishing House for their whole hearted co-operation in the publication of this book.

Author

CONTENTS

1

Introduction

Speaking at the World Economic Forum in Davos, Switzerland, on January 27, 2005, U.S. Senate Majority Leader William Frist stated that "The greatest existential threat we have in the world today is *biological weapons.*" He added the prediction that "an inevitable bioterror attack" would come "at some time in the next 10 years." He was seconded by Dr. Tara O'Toole, head of the Center for Biosecurity at the University of Pittsburgh: "*Bioterrorism* is one of the most pressing problems we have on the planet today." Are these statements realistic? Are they even proximately realistic?

By way of the most cursory comparison, one can set potential *bioterrorism* against:

1. Global climate change, which could affect populations in every corner of the globe, alter the current growth cycles of food crops that have evolved over millennia, and consequently food production;
2. The complex of global population growth, food production, energy and other resource constraints, and the waste products—solid, liquid and gaseous—produced by human society and the impact of these on regional and global ecosystems;
3. Ocean quality deterioration, deforestation, desertification, depletion of fresh-water aquifers—all of these are also global in impact;
4. Between 224.5 and 236 million people died in the 20th century in wars and conflict—say, roughly 230 million. This early in the 21st century, it is impossible to say whether the harvest of conflict-related deaths will be any different in the 21st century than it was in the 20th century.

5. If one adds deaths due to poverty, the figures become astronomical. Jeffrey Sachs currently estimates this sum worldwide at 20,000 people per day, or 7.3 million per year, approximately 75 million over a 10-year period. Some portion of the deaths that Sachs counts may be due to treatable disease: these are discussed separately below.
6. A working group convened by the Strategic Assessments Group of the Central Intelligence Agency (CIA) and the RAND Corporation in September 2004 listed 10 "future national security threats...to the United States" looking ahead to 2020. Of the 10, one was "proliferation of *weapons of mass destruction* (WMD)" and a second was "new health threats, such as *severe acute respiratory syndrome* (SARS)." There was no mention of the use of biological weapons by a terrorist group.
7. Turning to the 1999 Millennium Project list of "The 15 Global Challenges We Face at the Millennium," only 1 of the 15 dealt with disease agents: "What can be done to reduce the threat of new and reemerging diseases, and the increasing number of immune microorganisms." It did not include consideration of "*bioterrorism*" at all.
8. No attempt is made in this monograph to draw parallels— or to attempt a comparison of relative risk or potential consequences— between the prospect of "*bioterrorism*" and *cyberterrorism*. This is despite the fact that hundreds of attacks on Department of Defense (DoD) computers and on national infrastructure targets take place every day of the year, and that numerous successful penetrations have occurred. Available data a full decade ago indicated that as many as 250,000 attacks on DoD computers took place in 1995. A 1996 Government Accountability office (GAO) report characterized 65 percent of them as "successful."
9. Within a 10-day period between April 6 and April 16, 2005, no fewer than five other competitors were announced as being the most dire threat faced by nations:
   (i) nuclear terrorism;
   (ii) 640 million small arms and light weapons around the world, which are responsible for an estimated 300,000 deaths per year;
   (iii) a terrorist attack using high explosives aimed at cooling ponds holding stored irradiated nuclear reactor rods at civil nuclear

power plants, leading to reactor core meltdown and radiation release analogous to the Chernobyl reactor disaster;

(iv) the possibility of impact of an asteroid with the earth; and

(v) a missile attack that would detonate a nuclear explosive over the United States producing an *Electromagnetic Pulse* (EMP) "...that could come not only from terrorist organizations like al Queda but from rogue nations such as Iran or North Korea."

To complete the picture, Rogelio Pfirter, Director General of the United Nations Organization for the Prohibition of Chemical Weapons, stated, perhaps unsurprisingly, that "...*chemical terrorism* has been identified in different regions of the world as the number one potential threat."

If one looks only at disease and other human public health concerns, we see the following: Three diseases alone—*malaria*, *tuberculosis*, and *human immunodeficiency virus/acquired immune deficiency syndrome* (HIV/AIDS)—kill 5 million people globally year in, year out. In 2004, that sum reportedly reached 6 million. In 1 decade, that is 50 million dead. And although the contribution of HIV/AIDS to this sum is more recent, the overall total has apparently been roughly the same for many years past. Projected HIV/AIDS mortality estimates are available for the decade to come, and will very likely produce another 50 million dead due to these three diseases. *Falciparium malaria* is estimated to currently infect 515 million people worldwide, with 2.2 billion people—one out of every three people in the world—at risk of infection.

The cost of malaria to the economy of African nations alone is estimated at $12 billion per year. Malaria is preventable or treatable. *Tuberculosis* currently infects nearly 1 billion people, and 1 billion new cases are anticipated by 2020, 35 million of whom will die. The direct impact of global infectious disease was recognized by the Government in 1996 by the establishment of "...a national policy to address the threat of emerging infectious diseases through improved domestic and international surveillance, prevention, and response measures," and a standing Task Force in the *National Science and Technology Council* (NSTC) to coordinate those efforts, and by a January 2000 National Intelligence Estimate titled The global infectious disease threat and its implications for the United States.

1. Diarrheal diseases kill 3.5 million people per year. Most are preventable. The number can be considered more or less constant over many years, and would mean roughly 35 million dead over a period of 10 years.

2. When the Framework Convention on Tobacco Control—a global anti-tobacco treaty—came into effect on February 27, 2005, *World Health Organization* (WHO) authorities stated that smoking kills 13,500 people per day, or 5 million people per year. That is another 50 million people every 10 years. The world has an estimated 1.2 billion smokers, and notably the United States (and China) are among the treaty signatories that have, however, not ratified it. In this sum are 500,000 deaths per year due to smoking in European Union (EU) countries, and probably an additional half to two-thirds that number in Russia.
3. On March 4, 2005, WHO announced that measles mortality had dropped from 873,000 in 1999 to 530,000 in 2003. That amounts to 3.6 million dead in the past 5 years. As measles mortality was over 1 million children alone per year "as recently as a decade ago," that would mean around 10 million for the decade 1989 to 1999, and the same for preceding decades. (In early 2005, the *nongovernmental organization* (NGO) Doctors Without Borders, claimed that measles was still "killing nearly a million children every year.") Almost all nonimmune children will contract measles if exposed to the virus. The measles vaccine has been available for 40 years. Immunization costs U.S. $0.30 per vaccination.
4. Between the 22 years from 1977 to 1999, *flu* killed 788,000 people in the country, an average of 36,000 people per year. Even if there is no outbreak of *pandemic flu*, one can project another 360,000 *flu-related deaths* in the United States alone in the coming decade.
5. WHO officials have been warning for years of the imminence of a *pandemic flu* outbreak. They have warned that the pandemic was "almost inevitable," and long overdue. There were three influenza pandemics in the 20th century. The worst was the 1918-19 "*Spanish flu.*" Estimates of the mortality it caused range from a low of 20 million to higher estimates of over 50 million worldwide and even 100 million. The Asian flu of 1957-58 killed about 1 million people globally, while the third *flu pandemic*, the 1968-79 *Hong Kong flu*, killed an estimated 1-4 million people.

Flu pandemics usually occur about every 20 to 30 years, and it has now been 40 years since the last one. The H5N1 *avian flu* strain, which first appeared in the 1997 Hong Kong outbreak, is still believed to be generally incapable of person-to-person transmission. Until very recently, the human lethality of the H5N1 strain was considered to be

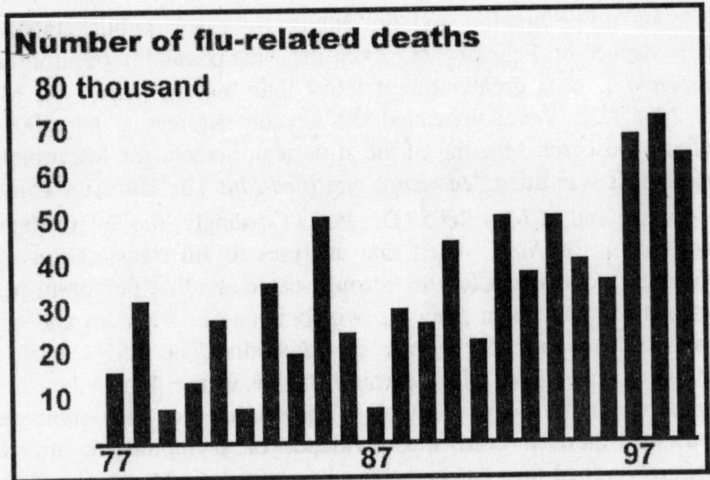

Fig. 1.1. Flu-related death.

approximately 50 percent of those infected. However, it is now understood that the actual incidence of infection is substantially higher, due to cases that produce atypical symptoms or do not show any symptoms at all, as well as because of underreporting from affected countries such as Laos, Cambodia, and China. While this happily means that lethality is actually lower than 50 percent, it also means that the chance of recombination to a human-transmissible variant is very much higher, since a much larger population is being infected.

What has been feared for the past half-dozen years is the recombination of the H5N1 *avian flu* in a person or animal simultaneously infected with a human flu strain, so that the recombinant would acquire the ability to be spread from person to person. Such a recombinant strain would display the lethality of the strain that caused the 1918 to 1921 flu pandemic which killed tens of millions worldwide. On February 23, 2005, the WHO Asia Director stated that "The world is now in the gravest possible danger of a pandemic." Dr. Julie Gerberding, head of the U.S. Centers for Disease Control and Prevention, told a meeting of the American Association for the Advancement of Science 2 days earlier that avian flu poses the single biggest threat to the world at present, the "most important threat that we are facing right now."

Even Dr. Anthony Fauci, Director of the National Institute for Allergy and Infectious Diseases, has finally discovered the seriousness of pandemic flu. Dr. Fauci, who in recent years made bioterrorism his main preoccupation and the subject of many public statements, acknowledged on February 28, 2005, that "the possibility of an influenza pandemic ... is a greater threat today than bioterrorism." On March 21, 2005, Dr. Fauci presented the keynote address at the 2005 Bio Defense Research Meeting of the American Society for Microbiology. His address was titled "*Influenza* and *Bird Flu*: The Ultimate Threats."

At the end of May 2005, Dr. Peter Cordingly, the WHO Regional Coordinator for Asia, stated that analyses of flu transmission in the preceding months had led to "strong suspicions that person-to-person transmission had taken place in two foci, one in Vietnam and one in Thailand," and possibly a third in Cambodia. The H5N1 strain was continuing to evolve. "It's learning to live in the human house, and that is potentially very worrying." The fatality rate also continues to drop, and there is confirmed evidence of asymptomatic infection. Cordingly added that on the basis of the timeline demonstrated in the 1957-58 outbreak, the pandemic could begin within 6 months of demonstration of sustained person-to-person transmission. In recent Senate testimony, Dr. Craig Venter suggested that "outbreaks and spread of avian and other flu virus strains ... could potentially kill hundreds of millions and wreak havoc with our global health system." Dr. Venter may have had advance notice of the report of the Global Task Force for Influenza, which appeared in the May 26, 2005, issue of *Nature*. This report predicted that 20 percent of the world's population would be infected in a *flu pandemic*, but estimated that perhaps only 7.5 million people would die. Writing in the special issue of *Foreign Affairs* in the summer of 2005, Dr. Michael Osterholm extrapolated estimates of the 1917-18 pandemic flu mortality to the current world population and arrived at 180 to 360 million deaths.

However, the U.S. Fiscal Year 2006 budget is to provide $4.2 billion to the Department of Health and Human Services for biodefense programs, $1.76 billion of which will go to NIH for biodefense research. At the same time, NIH will only be spending $120 million—less than one-tenth as much—for work on *influenza*. Imagine, in the current climate, what the reaction of the U.S. Administration, Congress, the media, and bioterrorism publicists would be if an agency were predicting with near certainty that "*bioterrorists*" were "about to launch a biological weapons attack that would kill somewhere between 10 and 100 million people, perhaps more."

1. On March 1, 2005, 758 microbiologists sent an open letter to NIH Director Elias Zerhouni criticizing NIH expenditure priorities

which greatly favoured research grant funds for the "*select agent*" pathogens of particular interest to biodefense. The supplementary materials that accompanied the letter included a brief discussion of the spread of antibiotic resistance to major human pathogens: The 2003 National Academy of Sciences report "*Microbial Threats to Health*" warned that "The world is facing an imminent crisis in the control of infectious diseases as the result of a gradual but steady increase in the resistance of a number of microbial agents to available therapeutic drugs," and recommended that "The U.S. Secretary of Health and Human Services should ensure the formulation and implementation of a national strategy for developing new antimicrobials."

These threats are posed by bacterial agents now established in human populations. *Tuberculosis* is in global resurgence. The World Health Organization projects that there will be more than 10 million new cases of tuberculosis in 2005 and that there will be nearly 1 billion new infected people by 2020, 200 million of whom will become seriously ill, and 35 million of whom will die. Additional threats are posed by other bacterial agents, including the agents responsible for *salmonellosis*, *shigellosis*, *borreliosis*, *legionellosis*, *ehrlichiosis*, *pertussis*, *syphilis*, *gonorrhea*, *chlamydia*, *meningococcal infections*, and *staphylococcal infections*. For each of these agents, strains resistant to multiple current antibiotics have emerged, and strains resistant to all current antibiotics have emerged or are expected soon to emerge.

In all of the above, which has been by way of introduction, two groups of global problems faced by humankind are listed. The second group enumerates only a few pathogen and public-health problems, with an annual mortality of above 11 million people per year. So, is bioterrorism "the greatest existential threat we have in the world today"? And is it "one of the most pressing problems we have on the planet today"? No. Absolutely not. That is clearly demonstrated by the above examples.

A 2003 report for the Century Foundation nevertheless noted, that the 2001 "*Amerithrax*" events demonstrated that "...bioterrorism could have an uncertain, far reaching, and potentially devastating impact." The statement that the release of a biological pathogen by a terrorist group should be considered as an occurrence of "low probability but high impact" is correct, but only with important qualifications. It does not mean any release of any agent formulation under any circumstances.

Rather, it presumes the release of a very high quality product, efficiently distributed under optimum conditions. Later sections of this monograph return to this question in more detail.

There were repeated statements in 1999, most prominently in the September 1999 GAO report, Combating Terrorism: Need for Comprehensive Threat and Risk Assessment of Chemical and Biological Attacks, that no threat analysis of this subject—an examination of specific potential actors, their capabilities and intentions, and potential feasibilities—had ever been prepared inside the U.S. Government. U.S. Government Accountability Office (GAO) reports for several years afterwards indicated that this situation had not changed. The task in this monograph is—in brief—to provide an evaluation of the overall problem, including several aspects rarely considered as being themselves contributions to the threat. This will be done by examining the following subjects:

1. The evolution of state biological weapons programs.
2. The evolution of nonstate actors ("*terrorist*") biological weapon capabilities.
3. Framing "the threat" and setting the agenda of public perceptions and policy prescriptions.
4. Costs and consequences of the U.S. biodefense research and development (R&D) program.

A threat assessment of the potential for the use of biological agents by terrorist groups is a very different exercise than that customarily faced in providing military threat assessments. No one ever did a threat assessment of a Soviet T-34 tank, the *Galosh antiballistic missile* (ABM) system that encircled Moscow, or an *Akula*-class nuclear attack submarine (SSN) without those systems actually existing. Declassification of historical intelligence estimates of the capabilities of forces and weapon systems that U.S. forces might have confronted have certainly provided examples of inaccurate evaluations. Nevertheless, performance characteristics and capabilities were often reasonably well known. Threat estimation of potential bioterrorism is as different as possibly can be from the assessment of a real operational system. It is almost purely hypothetical, and rarely, if ever, is predicated on a specific identifiable group and its capabilities. The range of possible assumptions is enormous, the utilization of extreme worst-case assumptions is the rule, and these universally depend on the projection of capabilities into the future, rather than their existence at the present time.

2

Biological Weapons

Biology today is as susceptible to hostile exploitation as were chemistry in World War I and physics in World War II. The formidable power of international commerce is behind this basic science, moving it toward innovations that, along with marketable medical value, might also be turned to destructive ends. If exploited by states, the science and technology of biological weapons could pose one of the most serious problems humanity has ever faced. A new generation of biological weapons, if pursued with vigor, could make them technologically competitive, especially for human control and domination. Unless the power of biotechnology is politically restrained, it could introduce scientific methods that would change the way war is waged and increase the means for victimizing civilians. The question at present is whether sufficient national and international restraints against this danger are in place, especially when scientific knowledge itself is at issue.

In the past, various, at times serendipitous, combinations of legal norms, public oversight, technical obstacles, and political leadership prevented the use of biological weapons. Overall, though, the world has been lucky, in that influential political actors took action at critical junctures and that the general historical trend of the last century has been toward transparency and open government. History shows that the problem of biological weapons proliferation is too complex to be solved by any single restraint.

A reasoned assessment of threats and measures is the first step to resolving the problem of proliferation. This problem should be understood as potentially more serious now than in the past, in great measure because human malice coupled with human ingenuity is a constant in world history. History also tells us that without a long-term commitment to nonproliferation, we are gambling with the future.

Trust and Mistrust

The United States' definition of its interests sets a standard to which the rest of the world cannot help but react. Even before September 11 and the 2001 anthrax letter attacks, the United States was in retreat from new international initiatives to strengthen the *Biological Weapons Convention* (BWC), while it also reinforced its homeland security policies. The Bush administration reinforced American unilateralism and a confrontational approach to international relations that, as in the Reagan era, depended on conspicuous military might. After September 11, "waging war on two fronts" was President Bush's apt description for America's militancy abroad and civil-defense orientation at home.

At the core of the US government's rejection of a strengthened BWC was the belief that the United States was exceptionally trustworthy and could therefore interpret and implement legal norms as it chose. Close allies of the United States, especially the United Kingdom, were trustworthy as well. In distinct contrast, suspect nations such as Iraq, North Korea, Libya, Syria, and Iran posed the most serious threat because they were too closed and perfidious to abide by law. Other states, such as China, India, Pakistan, and later Russia, were too well armed and large to ignore but still not fully trusted.

This perception of a global division between "*haves*" and "*have-nots*" is often shared among advanced industrial nations and has influenced treaty negotiations on weapons of mass destruction. From this viewpoint, the world appears divided into "*responsible*" Western states that can be trusted with nuclear weapons and "*irresponsible*" non-Western states that cannot be trusted with the weapons they have or that must be prevented from acquiring any.

In this view, arms control is but one element of a web of restraints imposed by the North on the South, encompassing export controls, strong biological and chemical defenses, and a "*determined and effective*" military response. A competing political vision assumes inseparable common interests among all the world's states in reducing the threat of weapons of mass destruction and of war and violence in general. It holds that the benefits of biomedical technology should be shared between wealthy nations and those in economic need. This perspective promotes the international mobilization of citizens, including scientists and physicians, to promote the interdiction of *biological weapons*. Far from being antagonistic to American or Western values, this approach is based on openness and democratic participation, extended to a global context. It encourages nongovernmental and grassroots organizations and the idea of civil society as part of the long-term international solution to present dangers.

The dichotomy between the two approaches is less sharp than it may appear. Despite the forces of globalization, the importance of sovereign state governments and state law in reducing weapons proliferation endures. The initiatives of advanced industrial states can be vitally constructive, provided there are options for broader cooperation. International measures based on trust are likewise essential for persuading a range of governments that openness through, for example, mutual onsite inspections and scientific partnership, is preferable to continued secrecy.

BWC and US Military Secrecy

At present 151 nations are parties to the Biological Weapons Convention, and 132 nations are parties to the Geneva Protocol, a strong testimony to the international norm. The BWC's lack of verification and compliance provisions is frequently referred to as its greatest shortcoming in reducing the threat of proliferation. What, then, is the solution?

US government objections to a protocol that would strengthen the treaty have been based on what it has argued is its exceptional need for military secrecy and the proprietary needs of its pharmaceutical industries. This double argument raises the question of whether, regarding this category of weapon, the safety of one nation is strictly divisible from all others, even, for example, from that of European allies like the United Kingdom, which have supported treaty compliance and transparency. The US defensive program is the largest in the world, a fact that US representatives have often underscored. But how large and, more important, how secret should it be to serve national interests—which are fundamentally public interests—or those of its allies?

The 1991 BWC Confidence-Building Measures, to which the US government agreed, require declarations of legitimate defensive projects and locales, leaving the substance of the classified work undisturbed. These declarations have compliance, not total transparency; as a goal. Further, they are intended as a means of distinguishing between legitimate programs and the illegitimate ventures that are more likely to be hidden and therefore undeclared. The United States should have no need for biological defense programs whose locations and general nature must be kept secret. But how much secrecy is too much?

If, as the covert CIA and Department of Defense projects suggest, the United States has a stake in undeclared projects at undeclared sites, it may be engaged in activities that increase the risks of proliferation and therefore danger to the public. With the world's largest military, the United States could set broad and dangerous standards for biological weapons research. Escalation within a defensive program, for example, could advance the laboratory and delivery technology for biological weapons. Small-scale, arbitrary projects, such as those leaked to the press in 2001, could pave the way for new agents or for attack simulations on a large, elaborate scale, to second-guess what a suspected enemy might develop. In time, other states or organizations would gain access to the same or similar classified technology; whether for a new agent, drug, bomb, or missile.

The US presumption that the threat of biological weapons is foreign is generally supportable, but several domestic crimes, including the 2001 anthrax letters, point to risks within the United States, perhaps from its own programs. Intelligence and defense officials may want to explore what an adversary might do, but secret knowledge and materials generated by their own biodefense projects could be put to hostile use by the increasing numbers of Americans with access and special skills.

Transparency in the name of public safety need not require front-page headlines, but it does require oversight and accountability. Ample US military funding without civil review fueled biological weapons proliferation in the past, and it could pose a danger again.

Legal Restraints and the Pharmaceutical Industry

The second US objection to strengthening the BWC concerns the protection of the proprietary rights of pharmaceutical companies. During negotiations on the BWC protocol, the lobby for the American pharmaceutical and biotechnology industry, Pharmaceutical Research and Manufacturers of America (PhRMA), was vocal in its objections to onsite inspections by international teams as an *economic threat*.

The cooperation of American pharmaceutical companies is vital to an effective BWC. At present their position resembles that of the American chemical industry in 1925 when the Geneva Protocol ratification was debated in the Senate. Competition then was strong among American companies and between American industry and European chemical manufacturers, while the non-Western world of vast colonies and emerging states lagged far behind in industrial capacity and consumption.

During the 1990s, pharmaceutical and biotechnology industries were be ginning to reap unprecedented rewards from years of basic research, much of it funded by the federal government. Earlier public fears about experiments with recombinant DNA had been resolved by the creation of institutional and federal government oversight committees. The exploitation of revolutionary technological advances, for example, gene-splicing and genome sequence databanks, promised to be highly profitable. In this competitive atmosphere, protecting patents and confidential business information was primary.

Predictably, the probusiness Bush administration defended the industry's right to secrecy. The pharmaceutical industry, though, is different from others in that its products directly promote health and save lives and therefore have a value that is universally subsidized by advanced industrialized nations for their citizens. That said, these corporations have responsibilities to stock holders and they protect their turf. For example, nearly every major pharmaceutical company joined in a 2001 suit against the government of South Africa's intent to import cheap generic AIDS medications, which the industry feared would set a global precedent. The bad press that followed this suit, which was subsequently dropped, motivated a number of the giant corporations to work with the World Health Organization to provide inexpensive, subsidized HIV medications for needy nations.

Nations differ greatly in the accountability they demand of their pharmaceutical industry and the controls they impose on them. Western European states, perhaps uniquely in the world, have required openness and account ability in the name of the public good. Perhaps as a consequence, Western European pharmaceutical companies were more accepting of protocol declaration and inspection proposals than PhRMA. In the United Kingdom, a strong supporter of the protocol, the political context for emerging biotechnologies allowed the public and environmentalists to review laboratory and manufacturing processes. This emphasis on social participation grew out of the government's assumption of responsibility for health and safety risks during World War II, which evolved into the United Kingdom's socialized medicine based on local citizen councils. In Germany, pharmaceutical accountability to the public has been accorded even higher priority, for the general environmental risks that manufacturers might cause. In contrast, the American approach to biotechnology has been to shield the manufacturing processes from oversight and to emphasize the product, which is judged by federal regulators and in the marketplace, where individual consumers might seek legal compensation if they have grievances. This American emphasis on product helps explain US industry's resistance to the BWC protocol.

Yet, throughout the years of Ad Hoc Group discussions, the emphasis was on inspection with regard to large-scale production, which left ample room for the protection of commercial secrets during sensitive research and development leading to patents. Even this end-of-process verification, though, proved unsatisfactory to the United States. With the US Bioshield program and increased funding for the National Institute of *Allergy and Infectious Diseases* (NIAID), American pharmaceutical companies now have a stake in civil defense that bears watching. The companies involved in biodefense are in a more precarious position than their officers may realize. They stand to gain only if there is a technical invention from basic research, for instance, a generally effective antiviral drug or other discovery beyond traditional defenses against the relatively arcane biological weapons agents. If pharmaceutical companies are drawn into national vaccine or drug distribution campaigns, they could emerge as heroes or, if there is a false alarm or exaggerated risk, they might be seen as purveyors of sickness and death.

Even before that, a spate of laboratory accidents, a product that causes injury in tests, a scientist who perpetrates a biocrime, extensive monkey and other animal research—any of these could harm the industry's image. In the long run, corporate leaders in biotechnology may eventually want to assume a larger role in preventing biological weapons proliferation, as the US chemical industry did after Vietnam and the use of chemicals in that war.

Preoccupied with defending American pharmaceuticals, the United States failed to consider how other nations—for example, China, Russia, India, and Brazil—with state-owned and state-protected biotechnology companies might be susceptible to military exploitation. The potential for "*niche states*" developing second, third, and further generations of biological weapons would be greater in the absence of verification and compliance measures. Iraq and South Africa are examples of program proliferation in an era of political indifference to their activities. Is such indifference supportable now?

Concept of Legal Completion

The United Kingdom, vigorously supporting the BWC protocol and legal restraints, argued that without pressure on all states to agree to verification measures (declarations and inspections) the opportunities for terrorists would increase worldwide: "*Improving states*' control of biological and toxin agents is a necessary component of international cooperation to ensure that they do not fall into the hands of terrorists." States have legal means to restrain potential proliferation, if they are called upon to use them. British scholar and BWC analyst Nicholas Sims has identified five legal treaty requirements approved by its review conferences but with which not all states that are party to the BWC have complied.

The first of these agreements concerns which calls on states to adopt national legislation to implement their obligations to the treaty. One such obligation is agreed to be the enactment of national penal legislation regarding activities that are prohibited by the Convention.

In 1991, states agreed to submit report annually the status of their compliance. National legal implementation has been uneven and slow (the United States passed its legislation in 1989) and, since it constitutes disregard of a BWC legal requirement, represents a threat to treaty compliance. A recent survey by VERTIC (Verification Research, Training and Information Centre) in London shows that percent of the 146 states parties have enacted an enabling law, with another 7 percent in the process of acquiring one. Africa was least in conformity, with only 16 percent of its nations having such laws, whereas in Europe, 73 percent of its states had passed the required legislation. Fifty-six states (37 percent of the world total) had no information to provide to VERTIC, nearly all of them nations in troubled world areas. Pakistan, Sudan, and North Korea were on this list, along with Saudi Arabia, Qatar, and Indonesia.

Second, a requirement to share information on national measures is related to national implementing legislation and also required. In communicating the means by which domestic legislation follows the BWC ban, a state demonstrates that it has translated the ban effectively and stringently, and it adds to a worldwide network of standard legal restraints.

Third, regarding the BWC, the first review conference declaration urged all states parties that had not already done so to become party to the Geneva Protocol. Thus far, twenty-five states parties to the BWC have not ratified or acceded to the Geneva Protocol. Again, most of them are in troubled regions of the world, the Middle East, Africa, Eastern Europe, and South America. In Sims's words, refusing to accept the Geneva Protocol sends "a message of equivocation" that weakens the BWC.

Fourth, the reservations to the Geneva Protocol that maintained an option for retaliation in kind—which permitted the old French and British programs—should have been withdrawn as states became party to the BWC. The terms are absolute: development, production, stockpiling, acquisition, and retention of biological and toxin weapons are forbidden "with the aim to exclude completely and forever the possibility of their use." About twenty state parties, including China, India, and North Korea, have retained reservations that might be understood as disdainful of the BWC.

Finally, the BWC calls for states to support negotiations to achieve a total ban on chemical weapons. Today, this is taken as an obligation to become party to the *Chemical Weapons Convention* (CWC), which is the result of those negotiations. Nevertheless, some dozen states party to the BWC have not joined the CWC.

International Criminalization of Chemical and Biological Weapons

The legal gap left by the various states' lack of national legislation in compliance is compounded by the lack of international law to criminalize biological and chemical weapons. Closing this loophole would make it difficult for a perpetrator to escape jurisdiction in a state's domestic court should that individual be found in a state that supports the law.

Beginning in 1996, a group of scientists and lawyers began formulating a new legal convention that would combine the prohibitions of the BWC and the CWC to make it a crime under international law for any person to develop, produce, acquire, stockpile, retain, or use chemical or biological weapons; to assist, encourage, induce, order, or direct anyone to engage in any such activities; or to threaten or engage in any preparations to use such weapons.

Seven international treaties already in force offer instructive models. Each provides the sort of expanded jurisdiction that could be incorporated into a new treaty for chemical and biological weapons crimes, so that it makes no difference in what country the crime was committed. Existing treaties such as the 1970 Airline Hijacking Convention, the 1979 Hostage-Taking Convention, and the 1984 convention Against Torture apply in any participating state in which the accused offender is found, even if the state has no links of territory with the offense or of nationality with the offender or with the victim or victims. Thus, in 1998, the membership of the United Kingdom in the Torture Convention made it possible for former President Augusto Pinochet of Chile to be detained in London for extradition to Spain.

Rather than identifying specific biological agents, the proposed criminalization treaty follows both the BWC and the CWC in their emphasis on hostile intent, that is, the purpose of the activity. In this way, technological change and the development of new agents and means of delivery are anticipated and the commercial use of many substances is protected. Provisions for dispute resolution and due process are part of the treaty proposal, as they are for similar treaties pertaining to crimes against humanity.

The prospects for this type of criminal treaty reflect the world climate for international justice. While the US administration is opposed to the establishment of the International Criminal Court, many other nations, including its close allies, have joined it. The prosecution of war crimes has a well established, if controversial, history from World War II. This history now includes tribunals established by the UN Security Council for prosecuting individuals for the mass murders that took place in Bosnia and Rwanda during the 1990 and in the former Yugoslavia in the 1980s. In 1998, the Rome Statute of the International Criminal Court introduced criminal penalties for the use in war of chemical weapons, using the phrase from the Geneva Protocol that bans "*asphyxiating*, poisonous or other gases, and all analogous liquids, materials or devices." It also points to state-level responsibility for infractions, defining such actions as those "committed as part of a plan or policy or as part of a large-scale commission of such crimes."

The fortification of legal restraints against biological weapons proliferation is bound to be a slow and difficult process of reinforcing norms in widely divergent contexts. Since the threat of biological weapons is global, then international legal initiatives are essential in order to protect US civilians and civilians everywhere, the primary targets of bioterrorism. Like all laws, they can succeed only if backed by legitimate authority and reinforced by other restraints.

Trust and International Control of Dangerous Pathogens

The control of dangerous biological pathogens to restrain proliferation has been approached in a variety of ways, often with an emphasis on trade restrictions. In the 1980s, in response to Iraq's use of chemical weapons against Iran, an organization of advanced industrial nations, now called the *Australia Group* (AG), assembled in Paris to agree on export controls on chemical weapons agents and their precursors and technologies. The AG later expanded its controls to include biological agents and equipment.

Through licensing measures on the export of certain chemicals, biological agents, and dual use equipment, the Australia Group's objective is to keep these items from contributing to the proliferation of chemical and biological weapons. By 2000, fifty-one pathogenic microbes and nineteen toxins were on the core list of agents subject to the group's export controls. The AG clearly divides the world into haves and have-nots. None of its thirty-four members falls in the category of poor developing nation; some excluded nations see threats to their economies in the Australia Group restrictions on dual-use technologies.

The United States has resorted to trade sanctions to control specific lesser states. In 1991, before its relations with Iraq deteriorated, the United States passed the Chemical and Biological Weapons Control Act, which promised that it would cease trade with any state that exported agents or equipment contributing to proliferation. This legislation specifically banned such ex ports to Iran, Libya, Syria, North Korea, and Cuba. In 1992 the US Congress passed the Iran-Iraq Arms Nonproliferation Act with sanctions against trade that could lead to WMD capability. In 2000, the Iran Nonproliferation Act specifically targeted Iran and sanctioned foreign persons trading in WMD technology, which included the Australia Group lists of agents and those of the 1993 Chemical Weapons Convention.

In 2002, the Bush administration approached controlling WMD technologies with a similar model, this time to encourage intelligence sharing among the major powers as well as the interdiction of shipments of WMD materials and equipment. On May 31, 2003, President Bush announced the *Proliferation Security Initiative* (PSI), a cooperative venture with nine European allies, along with Australia and Japan, to interdict illegal traffic relating to WMD at sea, in the air, or on land, and to streamline a rapid exchange of relevant intelligence. Within a few months, a series of simulations of interdiction raids began, by Australia in the Coral Sea, by Spain in the Western Mediterranean, and by France off its Mediterranean coast, and a tabletop air exercise by the United Kingdom. Described as an "activity" rather than an organization, PSI was intended to create "a web of counter proliferation partnerships through which proliferators will have difficulty carrying out their trade in WMD and missile-related technology."

The kind of trade control represented by the AG and the PSI may be less effective in relation to biological weapons agents and equipment than it is for either chemical or nuclear arms. Microbial pathogens, of which there are many in nature, can be transferred in minute amounts and later made to proliferate in quantity. The equipment needed to grow and process biological agents, as well as growth media, would be generic. Those in charge of raids would likely have preconceptions about violators based on their nationalities but might have difficulty discovering proof of culpability. Other proposals regarding the control of pathogens have avoided the thorny issue of trust among nations by focusing on scientific authority over research with dangerous pathogens. In one plan, tiers of scientific peer review could be built on institutional and national mechanisms already in place to control recombinant DNA research and research on laboratory animals and human subjects.

In the 1970, the US National Institutes of Health established the Recombinant DNA Advisory Committee (RAG) to review potentially hazardous research, with local research institutions gradually assuming control. Above such a national structure, an international organization, analogous to the WHO Advisory Committee on Variola Virus, which oversees US and Russian smallpox research, could monitor research. This plan and others for international monitoring of research are more directed at unintended biological hazards than at secret deliberate activities; but, by increasing transparency and creating international networks, they might also act to restrain clandestine efforts. The effective application of such monitoring to secret but presumably legitimate biodefense work would have to over come US opposition to any kind of international oversight of its activities in this area.

Homeland Security and National Science: Public Trust

During the BWC Protocol negotiations the Pentagon claimed that its role in defending the civilian population against bioterrorism justified the US requirement for secrecy. As the Department of Homeland Security (DHS) became established, it incorporated most Department of Defense programs related to domestic preparedness, as well as those from the Department of Energy and Department of Agriculture and elsewhere in the federal government.

The DHS is a new and experimental venture whose relations with other departments and agencies are still evolving. As its organizational culture takes shape, the greatest challenge to DHS is the cultivation of public trust, which is necessary to its mission to mobilize communities on their own be half. Trust in the department depends on its credibility in representing risks. If the DHS is secretive and adamantly closed to public inquiry, its account ability to civilian welfare can become suspect. Regarding accountability, DHS has inherited a domestic preparedness program from the 1990 that needs reevaluation and change. One problem is the way in which domestic preparedness restricts the public to highly reactive roles, as "*responders*" or as victims in fictive WMD attacks, instead of encouraging the initiative or feedback that serves best in responding to disaster events.

In addition, due to layoffs in local police and firefighters, the ranks of first responders have thinned. A state might have substantial federal funds for bioterrorism response training and many less police on the street. This change, even as it lowers routine security in communities, tips the safeguarding of the public to emergency federal management, which was never intended to make up the difference. What autonomy do states and cities have, then, in deciding what domestic preparedness and homeland security mean? Another problem looming for the DHS could be a bureaucratic loss of perspective on actual dangers, as crisis management becomes normalized. Political scientist Thomas Schelling, commenting on the 1941 Japanese at tack on Pearl Harbor, remarked that government agencies tend to foster "a poverty of expectations—a routine obsession with a few dangers that may be familiar rather than likely."

Since the September 11 attacks and the anthrax letters, federal officials have learned to communicate with each other despite agency differences in goals and methods. The creation of the DHS has accelerated that communication, with an increase in shared professional experience across the board. The pitfall may be that dedicated government officials concentrate on daily responsibilities in a rational environment, often ignoring unusual ideas or new frameworks for problem solving or perceiving risk.

Inside and outside DHS, federal officials now involved in homeland security must also reckon with intelligence data, who has access to it and who does not, and its interpretation. Secrets further complicate planning for local responses to unusual disasters and outbreaks of disease. The sharing of information up and down official lines of communication and across agencies is crucial in an emergency and should extend to the public. What, for example, does a FEMA (Federal Emergency Management Agency) official in DHS know that a local physician responding to an unusual outbreak does not? Or, to phrase the question differently, is there any government information about an unusual disease occurrence that should be withheld from citizens who must make decisions to protect themselves and their families? The answer is likely none at all.

The 1979 Sverdlovsk outbreak and the postal deaths from the 2001 anthrax letters both demonstrated the increased risks that secrecy and misinformation impose on the public. Future outbreaks may again show that information barriers cost lives while accurate communication reduces risks. Although largely ignored in current policy, the importance of basic, accessible, and good-quality health services, a combination of public health programs and clinical care, would be the best protection against an unusual deadly outbreak. This was Theodor Rosebury's argument in 1942, and it has merit today. Epidemics are opportunistic, targeting the more vulnerable populations. The better the general health levels in society, so that disparities are diminished, the less destructive the impact of a large disease outbreak would be.

There are other benefits to basic, accessible health care. In an emergency, people should know in advance where to go for protective medical interventions and how to get there quickly or make contact. For that, it helps greatly if they are already getting primary care from a local clinic, HMO, or hospital. They should also trust the medical personnel there who, in an emergency, would be in charge.

Similarly, informed participation at the local level is crucial in disaster situations and especially so with serious epidemics, which, to be defeated, demand public comprehension of how diseases are transmitted and what to do over the course of days or weeks. Before any unusual outbreak, the public should be educated about its central role in routinely preventing and containing infectious diseases and its members should understand disease risks and symptoms. No civil defense mobilization is required for this level of protection; good multilingual public health programs in community centers, clinics, and schools, via the mass media or electronically, would suffice. Without informed public cooperation, the American technological investment—in new laboratories and biological research, in biosensors, drug and vaccine stockpiles, and electronic tracking of disease—is nearly useless. The strictly national agenda of DHS to control bioterrorism is sure to be challenged. The risks of a serious US epidemic from any source would be difficult to separate from the risks and repercussions felt in other nations.

It is difficult, too, to predict which other nations, wealthy or poor or in between, will be affected by epidemics. For years, infectious-disease experts have warned of the international dangers of emerging infectious diseases, of which AIDS is the foremost example and the 2003 SARS (*Severe Acute Respiratory Syndrome*) outbreak among the most recent. As a recent Institute of Medicine report documents, the shared risks of serious outbreaks are more widespread than once thought. In an age of fast and frequent travel among nations, epizootics and epidemics travel across national boundaries, as witness the spread of West Nile virus across America, the annual influenza sea son, and the 2003 SARS epidemic.

As with large epidemics of the past, today's preventable epidemics of AIDS, malaria, tuberculosis, cholera, and other diseases, which kill millions every year, are part of the chronic cycles of poverty and violence that generate both despair and terrorism. This degradation of human life parallels the devaluation of human life on which total war and strategic weapons were based. Federal officials newly engaged in homeland security and biodefense may find themselves also contemplating the wider social reality of epidemics.

Biologists and Biodefense

Thus far, in confronting bioterrorism, federal policy has emphasized ingenious technological solutions. For example, DHS has instituted the Home land Security Advanced Research Projects Agency (HSARPA), modeled on a similar agency at the Department of Defense (DARPA) that funds extramural research on innovative defense technologies.

The greater part of the technology investment is in basic biomedical sciences, also for "*magic bullets*" to counter biological agents. In 2004, half the total federal budget for homeland security research and development (around $2 billion) was allocated to the National Institute for Allergy and Infectious Diseases for its own laboratories and for extramural projects to create pharmaceutical products. In this way, the civil defense goal to protect all Americans and a probusiness policy came together. The assumption on which this technology-driven policy was based is that, with only general reference to an actual bioterrorist threat, scientific research can defeat each established biological agent or their permutations or perhaps achieve defenses applicable to entire categories of diseases or toxins.

The US government's quest to incorporate biology and related sciences into its civil defense agenda heralds a significant change for scientists in these areas. As early as March 2002, the Defense Department was circulating proposed restrictions on the aims and conduct of scientific research, as a means of controlling the transfer of biological weapons technology. This reaction, fitting for other types of weapons, could barely begin to address the large, unwieldy field of biotechnology, which is as international as it is American. At one point, the FBI estimated to Congress that 22,000 US laboratories might harbor dangerous pathogens in old, leftover stocks from research on out breaks and epizootics. In theory, any scientist trained in biological laboratory techniques might undermine national security if determined to do so.

The federal requirements for laboratory and personnel registration written into the 2002 Bioterrorism Act were a comprehensive step toward controlling the potential threat. The same legislation implemented the funding of biologists to invent solutions to the bioterrorism threat, which would bring more scientific research under government supervision and align it with the larger military defense and probusiness framework for national security. Few US scientists in infectious disease research have much experience with secret government activities or with restrictions on the choice or substance of their work or the publication of its results.

Unlike nuclear physicists and engineers, biologists have intended exclusively beneficent ends to their research, ideally with universal application. Through peer-review processes, researchers in biology have been in charge of the distribution of generous federal funding since the end of World War II, with few strings attached. At times coercive federal politics did intrude. In the 1950s, for example, Elvin Kabat, coauthor of the Rosebury-Kabat report, was blacklisted after a research colleague alleged that he was a Communist.

As a result, despite his contribution to the war effort and consulting for the program that lasted until 1947, he lost federal research funds for several years. Scientific freedom from government intrusion (although not from government regulation) has been much more characteristic, as in the example of President Reagan's National Security Decision Directive 189, which affirmed that the products of fundamental research should remain unrestricted unless they were part of classified projects. This decision guaranteed scientists open review and acknowledgment of their research, while increasing the odds that government funding would go to top-level, creative performers. For biologists and other scientists engaged in biodefense research, decisions about the goals of their research are subject to the national security agenda, which gives priority to the interests of the United States over all other considerations.

The protection of American civilians, the stated biodefense goal, limits beneficence to one population and presumably withholds it from enemies, however broadly or narrowly they are defined. People in North Korea and Iran may be on that list. People in mainland China or Pakistan or other countries could be added, as political will determines.

After the Bioterrorism Act of 2002 and the infusion of new federal money into NIAID, the possible imposition of national security restrictions started a controversy involving DHS, the Department of Defense, the NIH, and scientific organizations, most notably the American Society for Microbiology (ASM). With billions of dollars per year newly directed to research on anthrax, smallpox, and other select agents, biologists have been asked to look at their research for the harm it might promote and the possible necessity to categorize it as classified or as "*sensitive but unclassified*."

If the parameters for evaluation were broadly set, national security threats might be read into many research articles, even those not intentionally related to biological weapons. A new laboratory technique, a new mechanical device, or an unforeseen discovery about a pathogen or drug could all be withheld from publication, to the detriment of the researchers, their institutions, and science in general. Predictably, the possible top-down federal censorship of research publications became a contentious issue. Should scientists instead self-censor their work, or should total openness prevail?

Contributing to this controversy were three original research articles, one regarding an engineered mousepox virus of enhanced virulence in laboratory animals, another on the production of an infectious poliovirus from laboratory reagents, and a third on enhancing the virulence of the vaccinia virus. To some minds, such research could inspire terrorists and should not have been published.

Among scientists, the opposing argument was that having the information made public increased the chances of countering potentially dangerous findings with more research and that this problem solving would happen faster, since more scientists would be involved. This debate took place with little questioning of why terrorists, who have had access to biological agents and small delivery technologies for decades, have rarely resorted to them. The ASM, which publishes eleven scientific journals, chose to monitor scientific research reports and instituted procedures for its reviewers to address potential national security risks in articles, primarily those on select agents.

In addition, in February 2003, the editors of major science journals, including Nature and Science, published a joint statement, "Scientific Publication and Security' to affirm their responsibility for monitoring submitted articles. They acknowledged that they might at times have to forbid publication of an article because of the potential "*harm*" it might cause, although the criteria for such censorship remained unformulated. Clamping down on potentially dangerous scientific knowledge promises to be difficult, if not impossible.

Ten thousand life science journals are published worldwide, the annual article submissions have recently reached nearly a half million per year, and as many as six thousand such journals are now available electronically, with assistance from the National Institutes of Health. American citizens are not necessarily in charge of all major journals or professional organizations, nor are they the majority of those who submit research articles. Intense and pervasive censorship would surely change this flow of information, with unpredictable consequences.

If research in biology might be used by terrorists or states to harm the public, what control can scientists collectively exert? A 2003 *National Research Council* (NRC) special committee report took the position that scientists should control judgments of what research constitutes a national security threat. Its major organizational proposal was to make the RAC the basic model for a layered system of oversight, to define "*Experiments of Concern*" that, like the development of a vaccine-resistant strain of pathogen, should be peer reviewed before being undertaken or published in full detail.

Individual scientists involved in biodefense research may have to wrestle with the elementary problem of retaining authority over the terms of openness for their projects. The policies, laws, and practices regarding classified and sensitive biodefense research can be interpreted in various ways and are sure to take years to evolve. In its report, the National Research Council committee exhorted, "Given the increased investments in biodefense research in the United States, it is imperative that the United States conduct its legitimate defensive activities in an open and transparent manner."

The two principles on which this call for openness is based are that the larger mission of science is to increase knowledge and that medicine is based on humanitarian principles that stand above nationality or other criteria. A third principle emanates from public health, namely, that the people deserve full, clear, and accurate information on disease risks and their protective options. On the other hand, since the goal of biodefense research is explicitly for national security, US military, intelligence, and homeland security agencies may legitimately impose secrecy on biological research or restrain the publication of its results. This restriction would be analogous to those already imposed on government scientists involved in nuclear and encryption research, which are integral to weapons programs. It is highly inadvisable that biology should become part of the defense industry, that is, have its Los Alamos or other secret government research enclave, because its goals are medical.

The international character of American science and medicine also stands in potential conflict with narrowly defined goals of national security. American biologists collaborate with colleagues worldwide and frequently attend and present their research at international meetings. Would this information sharing be curtailed by the US government? Within the United States, the biological sciences are international in their education of and reliance on foreign talent and expertise. A third of all US science degrees and half of the engineering degrees each year are awarded to foreign students, many of whom remain in the US workforce. US academic medicine has for decades been a training ground for graduates of overseas medical schools, who make up a third or more of public health and urban hospital staffs. In 2003, half the technicians at NIH were noncitizens.

In the aftermath of September 11 and the anthrax letters, federal concerns about educating foreigners in the sciences gradually took shape. President Bush's October 2002 Presidential Decision Directive (PDD-2), for example, allows the US government to "prohibit certain international students from receiving education and training in sensitive areas, including areas of study with direct application to the development and use of Weapons of Mass Destruction." The intent of the directive was the protective restriction of dangerous knowledge, but the determination of which students might be national security threats and the definition of the term "*sensitive areas*" may take years to clarify.

As with the 2003 smallpox vaccination campaign, large biodefense initiatives that appear risky to the public may have to be scaled down. Plans to increase Biosafety Level containment laboratories, initially ambitious, have encountered resistance from local communities in Davis,

California, Boston, Massachusetts, and Galveston, Texas, from those fearful of Sverdlovsk aerosol accidents or other disease disasters.

Biodefense research in general could have an important deterrent effect, in that no adversary would bother with an agent against which Americans might be well protected. But there are so many agents that the problem is hydra-headed. Biodefense research and Level laboratories unfortunately have the potential for increasing risks of exposure to the public. Therefore, the scale and substance of the entire enterprise needs careful, ongoing assessment. More research on select agents increases the known reserves of dangerous pathogens and therefore amplifies the risks of accidental exposure to laboratory workers as well as the risks of theft or biocrimes.

At NIAID, which dispenses its research funding nationwide, biodefense research started with standard biological weapons agents, such as those for anthrax, smallpox, tularemia, and viral hemorrhagic fevers. Research can extend to dozens more agents and to genetically engineered pathogens. NIAID director Anthony Fauci has advised that a whole new generation of biologists should be engaged in the field of biodefense research. But increased numbers of scientists trained to work specifically with biological agents may in itself pose a threat to civilian safety.

As discoveries move from the laboratory to final standardized products, more scientists and technicians will gain experience in animal and human testing of new vaccines and drugs and other products. This experience would presumably include tests and trials that simulate aerosol exposure during a biological weapons attack. Yet biodefense expansion proceeded with little consideration by scientists of its risks or why, after fifteen years of government warnings of bioterrorist attacks, none had occurred. Apparently federal funding for technological solutions to bioterrorism largely precluded political analysis and action.

Conclusion

When restraints on biological weapons break down, the risks of unprecedented danger increase. If history is our guide, the threat of biological weapons increases in direct proportion to government secrecy, closed military cultures, and a subsequent lack of accountability to the public.

Looking back at the last century, we can feel optimistic at the progress made thus far in deflecting the risks of biological weapons. The French returned to a biological warfare program after World

War II and then abandoned it. The British left off their program in the late 1950 and became leaders in treaty negotiations. The horrific program of the Japanese Army has had most of its secrets divulged and remains, with the Holocaust, one of the most distressing examples of barbarism within a great culture.

The US offensive program is no more; the giant Horton Test Sphere at Fort Detrick is on the National Register as a Maryland tourist site. The totalitarian Soviet Union, a political mastodon, took its offensive program to its historical grave. Now international consensus against biological weapons can be weighed against the complexities of the new wars born of globalization, poverty, and ethnic and religious fanaticism. In all, the strengthening of the Biological Weapons Convention offers the single best hope of global restraints. As before, the best protection against the proliferation of biological weapons is transparency.

Biological weapons are indeed a different class of threat, one that is variable in scale, in source, in its many potential agents, and in its possible impact on communities, including psychological and social disruption as well as illness and death. A radical argument can be made regarding both biological weapons and bioterrorism: that the policy goal should be an unobstructed flow of reliable information, in public education, in the review of scientific research, in disease reporting and diagnosis, in forensic investigation, in intelligence, in government oversight, and in international treaty compliance measures. Military and intelligence agencies may balk and pharmaceutical corporations may protest. It is for government leaders to show that their first priority is the protection of public lives, without qualification.

By virtue of its power and its democratic tradition, the United States bears a special responsibility to promote comprehensive, long-term restraints against biological weapons. Government secrecy, as the twentieth century shows, caused a degradation of biology and increased mortal risk to unsuspecting civilian targets.

The antidote to a reoccurrence of programs and threats worse than the ones of the past is the fostering of international relations that increase mutual trust and reduce the drastic economic inequalities that divide the world and obstruct communication. The United States is uniquely suited to lead this mission, if it can reconfigure its role in a politically complex world where national defense requires international cooperation. In all matters of disease transmission, freely shared information is essential to our protection. As Senator Daniel Patrick

Moynihan wrote in 1998, commenting on the end of years of Cold War secrecy and weapons proliferation, "Openness is now a singular, and singularly American advantage." It is an advantage that can be too easily lost, in short-sighted homeland security policies and in failed relations with other nations, to our peril and the peril of future generations. In military history, biological weapons were a failed innovation. May they remain so.

3

Biological Weapons Program

As World War II was ending, Secretary of War Henry Stimson ordered that research and development in biological warfare should continue. The Chief of the Chemical Warfare Service was put in charge of the program, and he was to be guided by the advice of the Surgeon General on medical aspects. The Army and Navy were expected to continue their cooperation. The newly established Air Force joined them in 1948. Partnerships with Canada and the United Kingdom also continued. For a brief time after the war, the US program showed considerable open ness about its wartime activities and their contributions to scientific research.

In October 1945, after arguing that an open publication policy would retain high-quality biologists and attract new ones, the Chemical Warfare Service was allowed to relax its publication rules. From then until 30 June 7947, 156 scientific papers passed approval at Camp Detrick. Of these, 121 were published in the open literature, 15 were accepted for journal publication, 20 articles were under review and another 28 papers had been presented at professional meetings. The subjects covered included aerobiology bacteriology, veterinary medicine, and the study of plant pathogens.

In May 1947, the formerly secret report by Rosebury and Kabat was published by the *Journal of Immunology*. Rosebury also brought out his book *Experimental Air-Borne Infection* as a publication of the Society of American Bacteriologists. This work contained a diagram of an aerosol chamber devised at Camp Detrick, photographs of equipment, and seventy-five tables of information on test results involving the anthrax simulant *B. globigii*, the *psittacosis* virus, *brucellosis*, and *tularemia*. Simultaneously, in stark contrast to this openness, biological weapons scientists from the *Chemical Corps* were involved in a secret arrangement to procure technical information from Japanese biological warfare scientists in return for protecting them from prosecution for war crimes. In Manchuria, from 1934 through 1945, the Japanese had conducted human-subjects experiments, including vivisection, on thousands of Chinese prisoners and had attacked and killed civilians with biological agents.

The immunity agreement, finalized in 1948, was a watershed event, a divide between retaliatory biological weapons as part of the World War II effort against the Axis powers and a new, Cold War vision of biological weapons as potentially comparable to the atomic bomb in destructive scope. The American bargain with the Japanese scientists helped reposition the US program as a top secret venture based on total war doctrine in the nuclear age.

Intelligence and the Immunity Bargain

In 1942, China presented the Allies with allegations that the Japanese Army had attacked its cities with *plague* infected fleas, causing significant deadly outbreaks. The Western press ran the story, which raised suspicions, but then the allegations faded from the news. With the defeat of Japan in 1945, the United States quickly sent scientists to investigate and appropriate innovative Japanese weapons technology, just as it had done earlier in Germany in what is known as the Alsos Mission.

The earliest scientific mission to Japan, in November 1945, included five civilian scientists and four members from the military, three from the Chemical Warfare Service. After six weeks of interviews and site visits, investigators found little of interest in Japan's chemical program and even less in its attempt to develop atomic weapons. One of the American conclusions, though, communicated to President Harry Truman, was that the Japanese biological weapons program might add to American military knowledge. Lt. Col. Murray Sanders, one of the CWS members of the group, quickly wrote a report summarizing Japanese accomplishments in bacteriology; this report was followed by another from the Army in 1946, written by Lt. Col. Arvo Thompson, which credited the Japanese program with developing effective vaccines and curative sera for plague, typhoid, dysentery, and other diseases, as well as diagnostic serums and antigens and drugs, including penicillin.

Japanese biological warfare scientists also reported having done research on animal diseases, including *anthrax* and *glanders*, and on *anticrop agents*. When questioned, the Japanese scientists denied to the American mission that they had prepared any *biological weapons* for offensive purposes or that they had engaged in experimentation on humans.

In early 1947, US military intelligence (G-2) agents in Tokyo invited Dr. Norbert Fell, a division chief from Camp Detrick, to reevaluate the Japanese program. He was not to estimate whether criminal prosecution was warranted, but to find out if its scientists had any valuable information that the United States should acquire for its national security. Gen. Ishii Shiro (1892-1959), the military biologist who had created the program, had been located at his home in the Tokyo suburbs; with him came the promise of important hidden documents being revealed. Might this information be worth some kind of negotiation? The same postwar question had been posed regarding Nazi Germany's weapons scientists, with whom deals had been struck for amnesty, immigration, and US defense employment. Those arrangements, as far as is known, did not involve war criminals who had engaged in brutal human experiments on captives, which the Japanese military scientists, despite their denials, had conducted.

G-2 officials appeared to leave the evaluation of scientific data, key to an immunity bargain, to Norbert Fell and several other CWS experts who also came to Japan. But by the time Fell arrived in Tokyo, in May 1947, US intelligence had already laid significant groundwork for a secret exchange. In late 1944, Secretary Stimson was told of the likelihood of a Japanese biological warfare program. In the following year, military intelligence gathered more supporting data, although the threat of biological weapons to the United States in the Pacific Theater seemed minimal. At the time, Washington was also interested in keeping the subject of biological weapons out of the news, lest the secret US program become public knowledge. In September 1945, the *Office of Strategic Services* (OSS), the forerunner of the CIA, concluded that the Japanese had "perhaps the best informed scientists in BW investigations of any nation in the world" and that they had posed a significant threat to the United States.

Concerning the Japanese scientific activities, a G-2 report from January 1947 commented: "Naturally, the results of these experiments are of the highest intelligence value." G-2 also managed to create a road block against the US Adjutant General's Office in Tokyo, which was tracking war criminals for prosecution. In March 1947, the Joint Chiefs of Staff sent an order placing G-2 in total charge of the biological warfare inquiry and mandating as essential "the utmost secrecy...in order to protect the interests of the United States and to guard against embarrassment."

To US leaders, the reconstruction of Japan as a stable democracy was crucial to offsetting the Soviet presence in Asia and the rise of Communist China. For Japan to become a credible ally, extreme Japanese aggression during the war, including the inhumane treatment of prisoners of war and civilians in the Pacific, had to be quickly adjudicated and forgotten. Many top Japanese military and civil leaders were already under indictment for war crimes. If rumors about the Japanese biological warfare program were true, the addition of prosecution for human experiments akin to those practiced in Nazi death camps would have added another obstacle to the goal of reconstruction.

In addition, Ishii's program might have had the patronage of the Japanese emperor, himself a biologist and known to take an interest in the details of government administration. The American government, and particularly Gen. Douglas MacArthur, who was in charge of the occupation and reconstruction of Japan, had decided that the divine persona of the emperor had to be conserved or the entire hierarchical structure of Japanese society could fall into chaos and perhaps cause violence against occupying troops. MacArthur, who controlled the war crimes trials, exempted the emperor from being prosecuted. Nevertheless, what the Japanese scientists might have revealed about his complicity in the biological warfare program could have been ruinous.

Entire networks of influence might have been implicated had the Japanese scientists been put on trial. Ishii had received support from the highest Japanese military and government officials. At least six of this cohort had already been indicted for prosecution in the Tokyo war crimes trials. Prime Minister Tojo Hideki, for example, when a high officer in the Kwantung Army (Japan's military force in Manchuria), had reportedly seen films of Ishii's human experiments; during the 1930s he had personally approved the expansion of the program. Gen. Umezu Yoshijiro, chief of the Army General Staff and a signatory to the 1945 instrument of surrender that ended the Pacific War, had been a Kwantung commanding officer familiar with Ishii's program. Officials not indicted had probably also supported the program.

Ishii was well connected in the medical world as well, which continually lent support to his program. Over the years, using those contacts, he had recruited hundreds of young medical students and technicians who circulated through its research facilities and were more or less complicit in the experiments. Sworn to secrecy at that time, after the war these educated people were poised to take positions in the new Japan's universities and industries. If the program were judged a war crime, blame might go beyond the military indictments to professional civilian networks, which would suffer guilt by association.

In May and June 1947, Norbert Fell conducted interviews with Ishii, the visionary of the Japanese biological weapons program. With a small cadre of his colleagues, Fell concluded that what the Japanese had to offer was worth the trade for immunity. Mainly the material was firsthand accounts of a large, long-term weapons program and documents of medical experiments on humans (including some eight thousand microscopic slides and detailed *autopsy* data). The autopsies included those of victims of inhalational anthrax, of special interest to the Americans. A supplement to Fell's report noted that "such information could not be obtained in our own laboratories because of scruples attached to human experimentation."

In Washington, the US government agencies that were consulted about the immunity bargain had slightly differing perspectives. State Department officials pointed to the risk of US government embarrassment should the agreement become public knowledge. In September 1946 the International Military Tribunal in Nuremberg had defined similar experiments by Nazi scientists as war crimes. G-2 and the Chemical Corps officials argued that the Japanese technical information was vital for national security and could not be made public at a trial or the Soviets would acquire it.

Indeed, the Soviets had asked for access to Ishii and the other scientists, which the US government granted in 1947, after coaching the Japanese to keep their responses minimal. The Washington consensus was that national security took priority over criminal prosecution. Apparently Gen. MacArthur personally intervened to keep the Japanese biological warfare scientists from exposure.

Exactly who authorized the immunity bargain remains unclear. Official approval should have come from the highest level of US government, from President Truman, Secretary of Defense James Forrestal, or Secretary of State George Marshall. A memo of 24 June

1947 by Fell identifies Truman and MacArthur, along with Alden Waitt, the Chemical Corps commander, as recommending the decision that "all information obtained in this investigation would be held in intelligence channels and not used for '*War Crimes*' pro grams."

By the time the agreement was signed, in March 1948, the Tokyo war crimes trials had concluded without revelations about the atrocities committed by Japan's scientists involved in the biological program. The cooperating Japanese scientists and many others who had participated in the program were now free. Except for Ishii, who was not in good health, the dozen or so who directly benefited from the agreement became medical school professors and administrators or pursued successful careers in research or industry. Not all the Japanese military involved in biological weapons were as fortunate as Ishii and his associates. In 1949, a dozen captured by the Soviet Union as they attempted to flee China were put on trial in Khabarovsk, one of four Siberian cities that Ishii had planned to attack late in the war.

The five days of testimony reportedly attracted large crowds and had to be broadcast by loud speakers outside the courthouse. The summary version of the Khabarovsk trial appeared in English in 1950. Although the Khabarovsk testimony described human experiments already known to American and British officials, the US and UK governments dismissed the proceedings as Stalinist propaganda. The defendants were imprisoned, and by 1956 most were repatriated to Japan.

To Western intelligence, the short prison sentences given the Japanese defendants indicated that the Soviets had contracted a bargain for information not entirely unlike the American one and for similar purposes. US military intelligence had acquired what seemed reliable information on past Soviet biological weapons activities from a German chemical weapons scientist. It suspected that the USSR had researched biological weapons during the 1930 and into the war years and could soon have the scientific and industrial resources to create a strategic threat.

Modernist Context: Manchuria

The social and political context for the Japanese biological warfare program was imperial expansion into China. On 19 September 1931, in what became known as the "*Manchurian incident*," the Kwantung Army seized the pretext of being attacked to begin its takeover of all key Manchurian cities along vital rail lines. Within six months, with help from jingoistic media and ultranationalist leaders, the Japanese public succumbed to "*war fever*" and began to define Manchuria as a "*lifeline*" essential to its national survival. In early 1932, the Japanese recreated Manchuria as Manchukuo, a puppet state, with Henry Pu Yi, the last Chinese emperor, as ruler. In 1933, Japan resigned from the League of Nations following worldwide protest.

Manchukuo was planned as a modern utopia with new cities based on science, culture, and industry. Japanese intellectuals and businessmen, seeking freedom from the old imperialism, took advantage of this social experiment, which they wanted to be "as modern as Soviet planning, Italian corporatism, German State socialism, and the American New Deal." Among other visionaries, the inventor of the Japanese corporatist state, Kishi Nobusuke, used Manchukuo as a test ground for fascist economic policies.

As its expansionist objectives were being fulfilled in Manchukuo, Japan faced a military threat from the Soviet Union, its major military adversary. Despite international agreements and a neutrality pact designed to minimize friction, the Japanese chronically chafed at sharing Asian borders with the USSR. In 1939 open warfare broke out between the two countries. In 1941, Japan and the Soviet Union concluded another neutrality pact, although the Japanese were still wary of the Soviets.

Outside Manchukuo, Chinese Nationalist forces were in conflict with Chinese Communists, and both governments in the civil war were hostile to Japan. In 1937 a border skirmish not unlike the Manchurian incident gave Japan the opportunity to enter the *China war*. Japan's military superiority was unquestionable. So too was its willingness to brutalize Chinese civilians, as in the 1937 Rape of Nanking. Japanese dominion soon extended along the eastern coast of China, south to Canton and beyond.

Before this war, while Manchukuo developed as a Japanese "*paradise*" with new cities and industries, tight, often ferocious military and police control was exerted over Chinese insurgents, guerrillas, and so-called bandits who were the source of "*experimental material*" for Ishii's program. The Kwantung Army had four non-Japanese populations to subordinate: the majority Han Chinese who had immigrated from the south, the indigenous Manchurians, Koreans imported as labour, and Russians and other European emigres. All of these groups were considered racially and culturally inferior, especially the Han Chinese. The army was rarely held responsible by Tokyo for its methods; using its military police, it kept order by its own frontier rules. Manchukuo was thus an optimal environment for a secret military program that could evade legal and moral accountability, attract scientists with the promise of good pay and research opportunities, and conduct human experiments on a despised civilian population.

Military Doctor as Visionary

As early as 1927, Ishii was making the case to his superiors for a biological weapons program. Part of his reasoning was that, after signing the 1925 Geneva Protocol, most nations in the world would forgo biological weapons, giving Japan (a signatory but not then a party to the treaty) the outsider's advantage in developing them. Western biological science, already promoted in Japanese universities and medical schools, promised to bring modern efficiency to the Japanese arsenal. Biological weapons required no very large industrial base and they were relatively easy to produce.

In 1928, Ishii went on a two-year world tour to pursue his interest in bacteriology. As a junior officer, he had achieved some fame for inventing a portable water-filtration system for mobile troops, even demonstrating this device to the emperor. His goal on his trip abroad was to learn more about the latest in Western biology; using diplomatic letters of introduction, he visited research laboratories in Germany, France, the Soviet Union, and America. On returning to Japan, Ishii was appointed professor of immunology at the prestigious Tokyo Army Medical School and promoted to major. In his absence, the 1929 economic crash had brought an end to Japanese liberalism and a return to isolationist politics. The military, in wresting control of the government, was turning to Nazi Germany and Fascist Italy as exemplars and allies.

As Ishii lobbied for a biological program, he enjoyed personal support from the highest military levels, especially among ultra-nationalists, and from the head of the Ministry of Public Health. Soon Ishii was given his own laboratory building at the medical school. There he drew a distinction between highly secret "A" research on offensive weapons and *"B" research* on defensive ones.

In 1932, his drive to do *"A" research* in Tokyo presented security problems, and Ishii was given permission to relocate in Manchukuo, where he had the full cooperation of the Kwantung Army. The difficulty of maintaining secrecy at his first installation, in the northern city of Harbin, led to the construction of a large walled compound one hundred kilometers to the south, the Zhong Ma Prison Camp. The camp was built to house between five hundred and one thousand inmates. Two years later, when a cohort of escaped prisoners blew up the facility,

Ishii's military superiors in Tokyo immediately agreed to fund a new and larger research compound. Generous support for the biological weapons program continued until the end of the war in 1945.

Ishii's new, heavily guarded enterprise was built on the site of a cluster of ten villages, known collectively as Ping Fan, twenty-four kilometers south of Harbin. When completed in 1939, it consisted of seventy concrete buildings, including laboratory and pharmaceutical facilities, three crematoria, several unheated outbuildings, and an airfield for Ishii's private fleet of biplanes. Intense secrecy surrounded the program. For the Japanese who worked there, mail was censored and travel was restricted. Unmarked trucks delivered material and equipment. In 1941, Ping Fan was renamed Unit 731. Anda, north of Harbin, was the Unit 731 outdoor field area for bomb tests, which also used human subjects.

In 1936 the Japanese signed the Anti-Comintern Pact with the Germans and Italians, thereby finally rejecting its old alliance with the British and their presence in Asia, where Japan intended to rule supreme. The year 1936 also marked the construction of Unit ,100, named the Kwantung Army Anti-Epizootic Protection of Horses Unit, headed by Wakamatsu Yujiro, a military veterinarian who was also granted immunity by the US government in 1948. Located six kilometers south of Changchun, the new capital of Manchukuo, Wakamatsu's ultrasecret facility was nearly the size of Ping Fan.

Unit 731 and Unit 100 were at the center of a system of research satellites that eventually extended to other Manchukuo and Chinese cities. Japanese academic institutions and hospitals were also involved, including the universities at Kyoto (where Ishii had gone to medical school) and Tokyo. From these and other institutions Ishii regularly recruited young scientists and technicians with promises of high wages and innovative experiments. In 1939, Ei Unit 1644, a satellite of Unit 73 was added to the program in Nanking. During World War II, as Japanese control expanded from the eastern coast of China into French Indochina, Siam, and Burma, other biological weapons satellites were established as outposts.

The extent to which any of these smaller centers engaged in human experiments, as opposed to basic laboratory research or vaccine production, remains unknown. Defensive work was also conducted at Unit 731. It manufactured twenty million doses of vaccine a year, most of it a typhus vaccine. Research on vaccines for *anthrax*, *plague*, *gas gangrene*, *tetanus*, *cholera*, and *dysentery* were also part of the program's agenda, along with research on blood products to treat a variety of diseases and for diagnosis. As for volume pathogen production, the Japanese lagged behind the industrial levels achieved, for example, by the Camp Detrick pilot plants for anthrax.

Unit 731, Unit 100, and the other sites in Manchuria and China were blown up by the Japanese before they fled in 1945. Documents were also destroyed, which made the information saved by Ishii and others that more valuable.

Japanese Biological Weapons Activities

Camp Detrick representatives were keenly interested in any mechanical delivery systems (bombs, artillery shells, or spray devices) that Ishii might have developed and in reports of actual biological attacks. The Americans ultimately found they had little to learn. The Japanese attempted but were unable to invent any bombs that were practical; although nine different types were developed, along with a prototype bomb for *anthrax spores*, called the *HA bomb*, none had effective dispersal capacity. Under Ishii's direction, over a period of five or six years, more than two thousand steel-walled bombs were detonated in field tests using humans who were bound in place at the site.

Many of the victims were killed by shrapnel wounds rather than biological pathogens. Ishii also experimented unsuccessfully with spray bacteria from low-flying planes. He had the intent but not the technology to link biological weapons to the airplane and strategic attack plans. Some joint experiments were held with the Japanese chemical program, apparently also involving human victims. But in general the chemical program was a separate venture, unlike in the West, where chemical weapons expertise and technology aided innovations in biological weapons.

Thus limited, Ishii devised technologically unsophisticated tactics that were more like disruptive sabotage. One was to lure enemy troops into a former battlefield secretly contaminated with infectious disease. Another was to inundate a civilian area with plague-infected fleas, either dropped by small low-flying planes or scattered by spies.

In 1939, Ishii was given the chance to use his bacteriological weapons. Throughout that summer, the Japanese attempted to gain control from the USSR of the Nomonhan border, a complex area near the frontiers of Mongolia, Inner Mongolia, Manchuria, and the Soviet Union. In August (a week after the USSR and Germany signed their

Non-Aggression Pact), Soviet Marshall Georgiy Zhukov led three divisions and five mechanized brigades to the border. There they encircled and routed the Japanese, who lost eleven thousand of their fifteen thousand soldiers in the conflict. As the Japanese retreated, a suicide detachment trained by Ishii stayed behind to infect the Khalkhin-Gol River with *typhus*, *paratyphus*, *cholera*, and *dysentery*. Whether these attempts had any effect on the Soviet soldiers was undetermined.

In 1940, Ishii organized five months of sporadic attacks on the southern port city of Ning Bo in Chekiang (Zhejiang) Province, in an attempt to disrupt food transports for Chiang Kai Shek (Jiang Jieshi), the leader of Nationalist China. Ishii used fleas to start plague outbreaks and attempted to infect water sources with typhoid and cholera germs. Japanese airplanes disseminated contaminated wheat and millet. Reportedly, a thousand people were made sick, with one hundred fatalities in the city and region. Ning Bo and nearby towns were affected by plague outbreaks the following year and again in 1946 and 1947. Some shipping and food transport may have been disrupted by these attempts, but under the general conditions of war, the Japanese found this effect difficult to measure.

In 1942 Ishii became involved in a difficult military campaign waged broadly through Chekiang Province. For example, in and around the cities of Yushan, Kinhwa, and Futsing, while the Japanese army staged a strategic retreat, three hundred of Ishii's men stayed behind to spray germs and infect water supplies, and three planes under fighter escort dropped bacteria and plague-infected fleas. Then, it was hoped, unsuspecting Chinese forces would invade the area and be overwhelmed by epidemics of cholera, dysentery, anthrax, plague, typhoid, and paratyphoid. Although Ishii later claimed that this attack was a great success, it was more likely a disaster. Ignorant of the germ attack, many Japanese soldiers apparently were exposed and as many as a thousand died. Shortly after, Ishii's authority was curtailed; he was removed as head of Unit 731 and transferred to the Nanking post.

The Japanese army also experimented with food and water contamination. In the Chekiang campaign, three thousand Chinese prisoners of war were given food contaminated with typhoid and paratyphoid bacteria and then released to spread the disease unknowingly as they returned home. This food distribution (but not its aftermath) was filmed and shown in Japan as an example of Kwantung Army beneficence to Chinese peasants. Japanese soldiers were also given contaminated food to leave behind along roadsides for hungry Chinese civilians or soldiers to find. In exchange for immunity, in addition to their laboratory records, Japanese biological warfare scientists reported to US military scientists and intelligence officials the details of twelve field "*trials*" in which they used biological weapons, including those just described, plus other plague attacks on Chinese cities and towns.

One advantage for the Japanese, which Ishii understood, was that in occupied territory and wartime China their germ weapon attacks could also be mistaken for natural outbreaks: This was one rationale for using plague-infected fleas. As the war was ending and Japan's position grew desperate, Ishii at tempted to convince top Japanese military officers to resort to biological warfare. His superiors generally resisted his plans. Yet earlier, in 1941, the Japanese army reportedly considered biological attacks on the Philippines in 1941, and in 1942 it listed Australia, India, and Samoa as potential targets, and then Burma and New Guinea.

The concept of total war was not foreign to Ishii. His chief mentor, the powerful Gen. Nagata Tetsuzan, was a strong advocate of military modernization with strategic objectives that included civilian targets. In 1947, after reviewing some sixty pages of Japanese material, the report by Detrick microbiologist Norbert Fell concluded that in the conduct of biological warfare the Japanese had little to offer, but he was optimistic about the human experimentation data: "It is evident that we were well ahead of the Japanese in production on a large scale, in meteorological research, and in practical munitions... However, data on human experiments, when we have correlated it with data we and our Allies have on animals, may prove invaluable, and the pathological studies and other information about human diseases may help materially in our attempts at developing really effective vaccines for *anthrax*, *plague*, and *glanders*."

Biological Weapons Agents: What was Learned?

Fell's optimism about the promised Japanese human experiment data must have been short-lived. Ishii, who controlled Unit 731 and Unit Ei 622, took a scattershot approach to investigating a fairly wide range of biological agents. Physician Edwin Hill, the Chief of Basic Sciences at Detrick sent to Tokyo in 1947, reported back to CWS commander Waitt a list of twenty diseases explored within the program, plus unspecified anticrop agents.

Ultimately, the Japanese scientists presented the Americans with autopsy material in twenty different disease categories with a total of 859 human cases. Of these, 3 were categorized by them as "adequate" for analysis. The majority of deaths were attributed to cholera (50), Korean or Songo hemorrhagic fever (52), tuberculosis (41) and bubonic plague (106). (The Japanese divided the plague autopsy cases into 64 "*epidemic*" and 42 laboratory cases.) As Fell wrote in his assessment: "In general it was concluded that the only two effective B.W. agents they had studied were anthrax (and this agent was considered mainly useful against livestock) and the *plague-infected flea*. The Japanese were not even satisfied with these agents because they thought it would be fairly easy to immunize against them"

It was under Dr. Wakamatsu at Unit 100 that these two diseases, anthrax and plague, were most thoroughly researched. Using animal research, US and UK scientists had developed an anthrax aerosol and considered it the most potent means of killing an enemy. Since the human inhalational form of the disease was rare, though, they kept returning to the problem of dose response and the difficulty of knowing how the disease progressed. On appearance, Wakamatsu and his colleagues were offering the Americans a rare prize: twenty-two autopsy records of human inhalational anthrax cases, among men from age twenty-five to thirty-seven, complete with colour drawings of the fatal progression of the disease.

The Camp Detrick scientists noted that such a small number of cases had no statistical value. Another obvious drawback was that the experimental conditions under which the anthrax victims were fatally infected represented nothing that would occur in battle or in mass attacks on civilians. Lacking a cloud chamber or techniques to create a fine anthrax aerosol, the Japanese resorted to putting four men at a time in a small glass chamber fitted with spray pumps operated from outside. High volumes of crude anthrax spore suspensions were then pumped directly at the victims' faces. The victims first developed tonsillitis and bronchial lesions, perhaps from sheer inundation with anthrax spores. After that, the infection spread quickly. The outcome for all was rapid death, two to four days after exposure.

Biological Warfare Scientists

The research activities of the Japanese biological warfare scientists had much in common with those of their contemporaries, Nazi Germany's biologists and physicians, who also conducted inhumane experiments on captives. The motivations for American biological warfare scientists to aid in protecting Japanese war criminals can only be guessed. They obviously placed national security above ethics and justice. Much later Murray Sanders said the immunity bargain was meant to protect the emperor's authority for the immediate purpose of preventing Japanese violence against occupying American soldiers.

The future of the US biological weapons program might also have been at stake. The postwar program could have been permanently reduced to defensive work with vaccines and masks, unless technological progress for offensive purposes could be imagined, against a substantial political enemy. Working in concert with G-2, the Detrick scientists were apprised of the Soviet threat and positioned themselves to respond. On 24 June 1947, Norbert Fell reported to military intelligence through his superior, Commander Waitt: "The information that has been received so far is proving of great interest here and will have a great deal of value in the future development of our program?' Fell then indicated that important data was forthcoming from the Japanese, for example, a report on mathematical calculations of aerosol dispersal.

Perhaps, too, the Camp Detrick scientists were covetous of a kind of forbidden fruit, the human research information that their scruples prevented them from acquiring. They could not have conceivably conducted the kinds of experiments on anthrax, tularemia, glanders, and other diseases that the Japanese had spent years performing. Nor had the Americans crossed the line to using biological weapons the way the Japanese had done, however primitive their methods. Due to the reliance of Japanese medicine on Western and especially German science, the Japanese and the American scientists had a professional culture in common, which might have promoted some empathy.

Despite the language barrier, the Japanese use of Western laboratory and autopsy techniques, materials, and descriptive scientific terminology proved easily comprehensible to their American counterparts. Japanese autopsy reports, for example, used German terminology characteristic of textbooks assigned in US medical schools. One of the shared scientific norms was a clinical detachment from the suffering and death of the human subjects. In the available US documents, American biological weapons experts appear to accept the Japanese human experimentation data without moral revulsion, as if it were clinical science.

More narrowly, Dr. Fell and other emissaries from Camp Detrick shared a particular wartime military culture with their Japanese counterparts, whose secret program aims had matched theirs. Perfecting biological bombs, deter mining infectious dose levels, developing battlefield and sabotage use of disease, planning mass attacks on civilians—both sides had pursued these goals in the interests of national defense. Professional deference was accorded the Japanese scientists, with whom the Americans from Detrick and G-2 had tea and dined. Japan was no longer the enemy; the enemy was the Soviet Union.

On 12 December 1947, Edwin Hill enthusiastically noted that the Japanese reports were a bargain, costing relatively little to retrieve and process, given the years the well-funded program endured. He added, "it is hoped that the individuals who voluntarily contributed this information will be spared embarrassment because of it and that every effort will be made to prevent this information from falling into other hands." To his report, Fell appended the collegial sentiment that he and the other American biological warfare scientists were looking forward to Ishii's memoirs on his military career. Yet he and the other American scientists must have realized that those memoirs, if honest, would have been confessions of criminality.

In the process of negotiation and information disclosure, Ishii and his cohort quickly realized that the Americans were no more interested in exposing the Japanese program than they were. The American government wanted to protect the emperor and keep information from the Soviet Union. It was perhaps equally important for the continuation of the Camp Detrick program to keep the American public in the dark.

The years 1945-47 marked a brief period of openness about US defensive wartime research on biological weapons, and that information release alone created more media attention than the Army wanted. The grotesque details of the Japanese program could well have tainted the Camp Detrick program and restrained the governments of the United States and the United Kingdom (which was privy to the Japanese bargain) from continuing their offensive projects. After protecting SPD during the war, US intelligence continued to protect the biological warfare program after the war and into the Cold War, when the CIA developed a client's interest in biological weapons agents.

In 1948, when the Japanese immunity bargain was completed, the head of the CWS and the US biological warfare scientists became complicit with US intelligence in sustaining secrecy around the criminal violations in the Japanese program. Together, US government officials and the military scientists involved in the Japanese immunity bargain placed national security; as it was interpreted at the time, above the law and above morality. Soon after the bargain was struck, as the Cold War intensified, secrecy descended on the US biological weapons program.

Secrecy Breakdown

After the war, it was known that the Japanese Imperial Army had established a biological weapons program. Yet details about the program—the human experimentation information and the treatment of prisoners—were not well known to the public, and for years the governments of Japan, the United States, and the United Kingdom denied them. The Japanese attacks on Chinese cities and towns were given short shrift, as if the Chinese were untrustworthy reporters of the events. Corroboration from Japanese scientists who had the evidence to prove the Chinese right would have corrected that perception, but their testimony and reports were kept secret.

Two decades after the Japanese biological warfare program terminated, attention to its criminal aspects began emerging sporadically in the media. In 1976 the Tokyo Broadcasting System televised a documentary, *A Bruise—Terror of the 732 Corps*, shown in Europe but not the United States. Five former biological warfare scientists revealed how they received immunity and how all their documents had been given to the United States. The story was carried by *The Washington Post* but caused no great sensation. In 1981, an article on the program in the *Bulletin of the Atomic Scientists* by publisher John Powell attracted wider media attention. Powell had obtained documents on the Japanese program from US government archives, and he also relied on the Khabarovsk trial transcripts. That same year, Tsuneishi Kei'ichi of Nagasaki University published *Kieta Saikinsen Butai* (in English, The *Germ Warfare Unit That Disappeared*), based on his archival and interview research. Tsuneishi later assisted two British journalists, Peter Williams and David Wallace, in research for a 1985 Independent Television feature, Unit 731—Did the Emperor Know?

Over the years, various Chinese and Japanese witnesses to the activities of the Japanese program made public statements. Some Japanese former scientists, in addition to confirming the worst of the human experiments, tried to explain the racist nationalism behind them. Said one, "I had already gotten to where I lacked pity. After all, we were implanted with a narrow racism, in the form of a belief in the superiority of the so-called Yamato [national Japanese] Race. We disparaged all other races... If we didn't have a feeling of racial superiority, we couldn't have done it.

Sheldon Harris, whose 1994 book Factories of *Death* deepened the investigation of Unit 731, pointed to a persistent US and Japanese reluctance to release documents relating to the program. In 2000 the

US Congress passed the Japanese Imperial Army Disclosure Act, requiring "full disclosure of classified records and documents in the possession of the US government regarding the activity of the Japanese Imperial Army during World War II' the Japanese government, with only weak laws supporting public access to documents, remained silent.

In 2000 the Japanese court that heard the case brought by the families of Chinese victims admitted that great wrongs had been committed, but asserted that compensation was impossible because the 1951 San Francisco Treaty had put an end to compensation claims against Japan. Meanwhile, international scholarly investigation of the Japanese program continued and new material has emerged, at times unexpectedly. For example, also in 2000, around 450 pounds of program documents left by retreating Japanese were unearthed from a bunker in Hailar, near the Soviet border.

Secrecy and the Moral Deficit

The immunity bargain had profound implications. For one, it prevented open international consideration of whether biological weapons programs, patently anticivilian, should be altogether banned. Policies regarding chemical and nuclear weapons were repeatedly exposed to public review and debate, in great measure because the consequences of their use were historically known. The effects of chemicals in trench warfare in World War I were well documented. After the bombings of Hiroshima and Nagasaki, judgments were possible about atomic weapons—their strategic worth, their deterrence value, and their danger to civilians.

The Japanese state program was the first evidence of modern biological weapons use, however crude, and of the scientific methods that, although taken to extremes, were consistent with the objective of strategic attack. In the absence of historical examples, public debate about these weapons failed to take place. In 1947, some members of the American public might have favoured keeping a biological weapons program and others protested. We will never know. The decision rested solely with a few military authorities and those in government who protected the program's secrecy and guaranteed its future expansion.

A legal and ethical deficit was also incurred, insofar as knowledge about Ishii's murderous program could have strengthened norms against state aggression on civilians. The Nuremberg trials and other criminal proceedings and the testimony of thousands of victims laid bare the structure and ideology of the Nazi state, with its industrial-age genocide in the name of racial purity. Once the world knew the consequences of Nazi Germany's racialist extermination policies, a moral lesson was learned and laws against genocide could be developed. Similarly, the realities of the devastation to Japanese civilians caused by US atomic bombs in 1945 contributed to restraints on nuclear weapons proliferation. In contrast, the Tokyo war crimes trials did not reveal how a civilized, technologically advanced society could for years conduct inhumane biological weapons experiments and attacks.

The 1948 immunity bargain forfeited the important chance for more open discussion of the menace of biological weapons to civilians and the need to strengthen international law and other constraints against them. The postwar history of legal restraints on biological weapons became profoundly troubled by government secrecy and a loss of public accountability, features that defined the United States program for the next two decades. By keeping the dreadful secrets of the Japanese biological warfare program, the US government lent legitimacy to offensive biological weapons aimed against civilian populations. John Powell described this consequence of the immunity bargain: "An '*insidious weapon*' in enemy hands, biological warfare (BW) was transformed into an acceptable and valuable military tool when added to the American arsenal."

The military value of biological weapons had still to be judged, but there was no question that its program would continue to have a secure niche in state government. While Fell and others were interviewing Japanese scientists, President Truman signed the 1947 National Security Act, a reorganization of the military services and intelligence organizations in reaction to the bombing of Pearl Harbor and the perceived threat of a Soviet-backed communist take over of the world. Central to this reform was the threatening prospect of the Soviet Union as a nuclear power. Nuclear weapons would set the standard for the next twenty years of biological weapons development, making it imperative for biological warfare scientists to show how pathogens could devastate populations at the same enormous scale.

4

Evolution of Biological Weapons

Information derived solely from official U.S., Russian, and United Kingdom (UK) sources has been available since 1988 which specifies how many and which nations maintain offensive biological weapon programs. Official U.S. Government statements repeated for many years that there had been four nations in possession of offensive biological weapons programs in 1972 at the time of the signing of the Biological and Toxin Weapon Convention (BTWC), and that this number had increased to ten by 1989. In November 1997, the Director of the U.S. Arms Control and Disarmament Agency (ACDA), in the course of a statement to BTWC negotiating states in Geneva, increased the U.S. estimate to 12 nations. The additional two states have never been identified by U.S. officials. In July 2001, a U.S. Government official stated that 13 countries had offensive biological weapons (BW) programs.

There has been no equivalent statement or revised estimate since. All through the 1990s, it was common—in fact, nearly universal—for commentators to depict the proliferation of state BW programs as a constantly increasing trend. For example, Ambassador Donald Mahley, the senior U.S. diplomat to all multilateral negotiations in Geneva concerning chemical and biological weapons, stated in an October 1996 Voice of America broadcast that "It is estimated that over the last several years, the number of countries suspected of having a biological weapons capability has risen." However, it seems very possible that it may have been a more or less stable constant for the last 20 years. Moreover there were several notable reductions or deletions from the list in the past decade. Since the overall total was not very large to begin with, that would be a very significant reduction:

1. The BW program of South Africa was terminated by 1995.
2. On November 1, 2002, U.S. Undersecretary of State John Bolton stated that "Libya has an offensive BW program in the research and development stage, and it may currently be capable of producing small quantities of biological agents." The statement was consistent with other U.S. statements regarding Libya and BW during the preceding decade. The phrasing in the 1993 report of the Russian Foreign Intelligence Service was substantially stronger, stating that "There is information that Libya is engaged in initial testing in the area of *biological weapons*." At the end of 2003, U.S. and UK government teams working in Libya ascertained that Libya had never had an offensive BW program. In the words of a U.S. administration briefer, "Libya acknowledged past intentions to acquire equipment and develop capabilities related to Biological Weapons." Libya additionally "committed not to pursue a biological weapons program and to accept the necessary inspections and monitoring to verify that understanding." Apparently Libya may at some point have either procured or investigated the procurement of dual-use equipment that might have served such a program, information which had been picked up by intelligence. This experience demonstrates a weakness in judgments based on procurement monitoring. It can be a useful indicator, but it cannot be considered definitive.
3. It is now clear that under the pressure of UNSCOM inspections, the BW program of Iraq was disbanded between 1992 and 1995. Unfortunately, the Iraqi program provided two other lessons. First, as shown by the period 1985 to 1990, an offensive BW program can be hidden for quite a number of years, including during the period in which it initiates production. Second, as demonstrated in precisely the opposite direction by the period 1998 to 2002, the most basic errors in judgment can be made by Western intelligence agencies. In addition, further public political manipulation of those mistaken judgments by political elites can take place.
4. In 2004, the present U.S. administration also withdrew the charge that Cuba maintained an offensive BW program.
5. Given the continued total denial by the Russian government of international access to the BW facilities of the Ministry of Defense of Russia, as well as continued impeded access to relevant facilities

of the Ministry of Health, it is impossible to be certain of the status of BW-related activities in Russia. Nevertheless, they certainly are very greatly reduced from what they were up to 1991-92.

This would mean an absolute reduction by four states—South Africa, Libya, Iraq, and Cuba—roughly one-third or one-fourth of the total number of states that, according to the U.S. Government, maintain offensive BW programs. In addition, the status of several of the other national BW programs appears to be less certain than previously implied. If one looks at a recent public Central Intelligence Agency (CIA) assessment of the BW programs of Iran, North Korea, and Syria, it reads as follows:

Iran

Even though Iran is part of the BTWC, Tehran probably maintained an offensive BW program. Iran continued to seek dual-use biotechnical materials, equipment, and expertise that could be used in Tehran's BW program. Iran probably has the capability to produce at least small quantities of BW agents.

North Korea

North Korea has acceded to the BTWC but nonetheless has pursued BW capabilities since the 1960s. Pyongyang acquired dual-use biotechnical equipment, supplies, and reagents that could be used to support North Korea's BW program. North Korea is believed to possess a munitions production infrastructure that would have allowed it to weaponize BW agents and may have some such weapons available for use.

Syria

Syria probably also continued to develop a BW capability. A sentence on North Korea by CIA Director Porter Goss on March 17, 2005, was somewhat stronger: "We believe North Korea has active chemical weapons (CW) and BW programs and probably has chemical and possibly biological weapons ready for use." By political agreement among the States Parties to the BTWC, Confidence Building Measures (CBM) were to be submitted annually, beginning in 1987. Iran did not submit any until 1998 and 1999, and when they did, they conveniently "forgot" to submit perhaps the two most critical CBM forms out of eight. These were the declarations that require the state to list national biological defense research and development programs and past activities in offensive/defense biological research and development programs. In 2002, Iran declared that it "did not and does not have any national, subnational or individual programs/activities and/or facilities related to *biological offensive purposes*" and that it "did not and does not have any 'National Biological Defensive Program'.

However the state has carried out some defensive studies on identification, decontamination, protection, and treatment against some agents and toxins." Official U.S. statements regarding the Iranian BW program from 2001 onward reduced its apparent status compared to assessments that had been offered in the late 1990s. Reference to agent and weapon stocks disappeared. During President George W. Bush's first term, the administration submitted only one Arms Control Compliance report to Congress, although this report is intended by Congress to be an annual submission. It did this in its first year in office, and there has not been another one since.

It therefore remains to be seen how the Iranian BW program will be described in any forthcoming version. It is useful to recall a statement in 1999 by Dr. John A. Lauder, then the Special Assistant to the Director of Central Intelligence for Nonproliferation: *Intelligence* is all about ascertaining not only the capabilities, but also the intentions of one's adversaries. Because of the dual utility of the technology and expertise involved, the actual CBW threat is in fact directly tied to intentions. Getting at this intent is the hardest thing for intelligence to do, but it is essential if we are to determine with certainty the scope and nature of the global biological and chemical weapons threat.

U.S. Government officials have never explained what the word "*capability*" means in these statements: whether it means the procurement of dual-use biotechnology equipment, a national pharmaceutical production capacity, a dedicated defensive BW R&D program, or the identification of dedicated infrastructure for offensive BW R&D. It is clear however that after the Iraq *weapons of mass destruction* (WMD) intelligence failures demonstrated by the reports of the Iraq Survey Group, the U.S. administration decided to be much more cautious in the conclusions that it drew from perhaps rather ambiguous information. Now and then over the previous years, one had overheard a comment by a government official or former government official to the effect that the evidence regarding country X or Y was ambiguous or weak. But that body of relevant evidence was never available for examination.

In 2003, a WMDwide review of U.S. assessments was initiated, and the review on the proliferation of BW was headed by Lawrence Gershwin, National Intelligence Council officer for Science and

Technology. This apparently led to the readjustment of some previous assessments. In the case of Iran, an unusual opportunity was provided in 2003 by the presentation in Washington, DC, of detailed allegations regarding Iran's alleged BW program.

The information was provided by the same group that, in some cases, has been the first to provide information on *Iran's nuclear weapon complex*, information that had not been publicly known, nor apparently known by the *International Atomic Energy Agency* (IAEA), and which led to subsequent international action which forced Iranian government disclosures. The BW information was quite detailed, naming individuals, institutions, facilities, and locations. However, there has never been any comment or corroboration of these allegations by the U.S. Government or by any other government or international agency. In November 2004, CIA Director Goss reported to Congress that Iran continued "to vigorously pursue indigenous programs to produce nuclear, chemical and biological weapons." However, in March 2005, a special Presidential panel decried the state of knowledge available to the U.S. Government regarding even Iran's nuclear weapon program. The U.S. Senate Select Committee on Intelligence is currently reviewing the information available to the administration regarding the *nuclear*, *chemical*, and *biological programs* of Iran and North Korea. It will be interesting to see if that report is made public, and, if so, what it will say in regard to the BW programs of these two countries.

The CIA document containing the assessments of the Iranian, North Korean, and Syrian BW programs quoted above is released twice a year. Although the analogous paragraphs of the preceding half-dozen or so years have not been included here for comparison, the above statements are all lower key than they were in earlier years in the same report. The caveats are notable: "*probably*," "continued to seek," "the *capability to*," "would have allowed it to," "probably also continued to develop." No definitive statements of production, stockpiling, or the nature of munitions are included.

As always, there is no discussion of Israel's BW "*capability*" or the status of its BW program in any public U.S. Government report. It is interesting to note, in regard to the comments on proliferation that follow below, that in the only relevant data that appears to be available, Israel had the second largest number of visitors to the U.S. Army Medical Research Institute for *Infectious Diseases* (USAMRIID) following the UK, a country with which the United States shares BW relevant information.

We know that there was exaggeration of chemical weapon proliferation in the past, so this is not a unique experience. In 1990, Brad Roberts wrote "We entered the 1980s with three, four, or five chemically armed states; we will enter the 1990s with *upward of two dozen chemically armed states*". The estimate of the Director of the U.S. Arms Control and Disarmament Agency had been somewhat less, saying that "at least 15 states possessed chemical weapons, with others attempting to acquire them." However, on January 24, 1989, his successor, General William Burns, told the U.S. Senate Foreign Relations Committee that only five or six of these countries actually possessed stockpiles of chemical weapons in addition to the U.S. and the Union of Soviet Socialist Republics (USSR).

In August 2005, the U.S. Department of State released its most recent version of its "*Noncompliance*" Report. Although nominally an annual report, none had been released in 2004. Eight countries are discussed in the section dealing with the Biological and Toxin Weapons Convention. A sentence in the opening introductory paragraph of the section states that "the specific cases addressed here are those that have assumed obligations relevent to the BWC and for which the most evidence exists of actual or potential noncompliance." The insertion of the word "*potential*" would seem to substantially undercut the utility of the entire section. A country should be considered in compliance with the BWC, or not in compliance. Libya is still included as one of the eight nations, although previous evidence presented by the current administration and discussed earlier would provide no reason for that. Cuba also is still present.

The discussion of several countries was larded with caveats, explicit or implicit. Most problematic was the inclusion of general descriptors as suggestive evidence of noncompliance, such as work in "*advanced biotechnology techniques*," "*aerosolization techniques*," "legitimate public health and commercial uses [that] could also offer access to...BW enabling capabilities," "the technical capability to conduct limited offensive research," the existence of facilities that "could easily hide...capabilities for a potential BW program." All of these nonspecific indicators could equally apply to every NATO ally of the United States, every EU country, to many advanced developing nations, and above all, they would be more applicable to the United States than to any other country. They are not evidence of BWC treaty noncompliance. The use of such descriptors will almost certainly place previous U.S. attributions of BWC noncompliance in question in the minds of international diplomats who have regularly read the analogous BW sections in earlier annual U.S. Noncompliance reports. Finally, the 2005 report reinforces the conclusion that the trend of BW proliferation has not been increasing over the past decades, but that it has either been essentially constant or has been decreasing.

The possibility of proliferation from three of the former or present programs—South Africa, Iraq, and Russia—has been raised at times by commentators. What is known is as follows.

South Africa

The South African BW program was minimal, no more than a handful of researchers were involved. Contrary to various press reports in the media, the program did not include genetic engineering of pathogens, nor—as best is known—has there been any proliferation from the program whatsoever.

Iraq

There was no known emigration of researchers from the former BW program of Iraq. However, the Addendum to the report of the Iraq Survey Group released in March 2005 states that "Migration of some WMD-associated program personnel to countries like Iran or Syria is possible." The phrasing is ambiguous in that it cannot be deciphered if this is assumed to already have happened, or if it is a generic statements suggesting that such migration could take place in the future. No disaggregation of "WMD" is provided. The report continues: "Since OIF [Operation Iraqu Feedom], the ISG [Iraq Survey Group] is aware of only one scientist associated with Iraq's pre-1991 WMD program assisting terrorists or insurgents. However, there are multiple reports of Iraqis with general chemical or biological expertise helping insurgents produce chemical and biological agents." Nothing further is said about these "reports," or whether the ISG considered them credible or not. Much more significantly, the Addendum then discusses a list of Iraqi BW scientists prepared in early 2002 for the purpose of possible transmission to Syria. However, it was not known if the list was or was not ever transmitted to Syria, and the ISG did not discover any evidence of emigration of Iraqi BW scientists to Syria.

The *Twentieth Quarterly Report* of the United Nations Monitoring, Verification, and Inspection Commission (UNMOVIC) to the UN Security Council does raise the possibility that some of the pathogen cultures used for Iraq's BW production program may still remain unlocated in Iraq, a consideration that was also raised in the report of the U.S. Iraq Survey Group.

Russia

Again, contrary to undocumented hints that find their way into media reports, there is no known evidence of the transfer of pathogens from the Soviet or continuing Russian programs to any other state, either prior to 1992 or since 1992. Emigration of former Soviet BW-related researchers to any proliferant state has been minimal. The one known example is the move of 10-12 researchers, largely from the institutes belonging to the Russian Academy of Sciences rather than from former BW institutes, to Iran. There continue to be statements, particularly by Dr. D. A. Henderson and the Pittsburgh group, to the effect that the location of Soviet BW "*stockpiles*"—not culture collections—produced in the USSR prior to 1990 are not known. U.S. intelligence and defense agencies have believed for over a decade that those "*stockpiles*" were destroyed by the USSR roughly between 1988 and 1990, and there are no indications that these agencies have ever altered that judgment. Contrary statements appear to be deliberately misleading.

As late as June 2005 at a seminar at the Council on Foreign Relations in New York City, Dr. Henderson said in reference to the possible dispersion of smallpox from the USSR or Russia: There have been economic problems in the Soviet Union, or now Russia. Many of the scientists have left the laboratories. They've gone all over the world, different places, some in the United States, Europe, some have gone to North Korea, Iraq, Iran. So that [is what] the problem is, there is just no way of knowing who has what and where, and that's the concern, that there may be others with the virus, but we just can't find out about it.

No Soviet or Russian BW scientists are known by the U.S. intelligence community to have gone to North Korea or to Iraq. Those few Russian scientists that went to Iran did not come from institutes that worked with smallpox. The U.S. intelligence community does not believe that smallpox virus was transferred from the USSR or Russia to any other state, not the three mentioned nor any other. The great majority of Russian BW relevant scientists that emigrated did so to the United States, the UK, Germany, France, Australia, Sweden, Finland, Israel, etc., and not to states of BW proliferation concern.

Soviet Biological Weapons Program

The 1972 Biological Weapons Convention completed the codified restraints on germ weapons: not only their use but also any preparations or possession for such use were outlawed. Biological weapons, the only weapon of mass destruction then internationally banned, became legally distinct from chemical weapons, which had a demonstrated if limited tactical potential and some potential for deterrence, as during World War II when neither the Allies nor Germany dared initiate first use. Biological weapons also became legally set apart from nuclear weapons, which had known strategic power and the potential for deterrence against use during the *Cold War*. After 1975, the dominant issue for those concerned about biological weapons was how and if legal restraints actually prevented secret proliferation. Did any nation have the right to secrecy in the face of strong international consensus against germ weapons?

Problem of Promoting Compliance

After World War II, the severe military restrictions on Germany and the other Axis powers included bans on biological, chemical, and nuclear arms. With the 1972 *Biological Weapons Convention* (BWC), sovereign nations agreed to the ban. Yet, formulated without measures to promote transparency, the treaty could not prevent or reveal secret offensive programs if a state party was willing to risk violation. The BWC lacked strong compliance provisions, such as mandatory declarations of past and current capabilities that might be verified, a stable independent organization to oversee and encourage state compliance, and joint exercises or onsite inspections to foster communication among states and build confidence that the prohibition was effective. In the event of an unusual disease outbreak, no standard means were in place for credible expert inquiry. Allegations or complaints might be lodged with the United Nations Security Council, but there *Cold War* politics were likely to diminish the chance of disinterested inquiry, as they did in 1952 when North Korea accused the US of biological weapons use.

From 1969 to 1975, the United States set the maximum standard for compliance. It fostered high-level civilian review of its military program and policies, unilaterally repudiated biological weapons, and publicly dismantled its large offensive facilities. The United Kingdom also openly dispensed with its biowarfare program; by 1968 it was holding "*open days*" for the public at Porton Down to emphasize its public health mission. By 1968, France, a nuclear power, had also repudiated its postwar investment in offensive biological weapons and gave its support to the BWC. The Soviet Union, with the United States and the United Kingdom, had formulated the treaty's provisions and was with them a depository nation. The price of Soviet participation was minimal incursions on state and military secrecy. After signing the treaty, the Soviet Union fully exploited this weakness by covertly developing an offensive program that exceeded the capacity the United States had earlier achieved.

As the Soviet Union was collapsing, the secrecy surrounding its biological program broke down. Former program scientists began revealing details of a secret, bureaucratically enormous program, which employed thousands of scientists and technicians and supported large research and production facilities. Like the British and US programs, the Soviet program emphasized experiments on infectious agents and toxins and aimed for large-scale industrial production for strategic attack. Its technical achievements were uneven but intimidating. Without great success, it emulated Western science to increase the power of some germ agents. With time, it sought to add long-range intercontinental missiles to aerial delivery.

Beginning in 1991, exchange site visits involving Soviet, British, and American experts shed more light on the scientific and industrial dimensions of the USSR program. Later, other cooperative ventures added detail to the picture of the large state bureaucracy in which the Soviet program originated and grew, known to relatively few people even within the USSR. It took regime change to bring transparency to the Soviet system, so that the program was basically defunct by the time it was revealed. Nonetheless, the 1990 were marked by increased international experience in onsite inspections and documentation that pointed the way to improved verification and compliance. The most important question the Soviet program raised was how international law could diminish the biological weapons threat posed by a secret state.

The objective of the 1972 Biological Weapons Convention was and is to eliminate the offensive state-sponsored programs and their dangerous products. A major problem in assessing state compliance is the secrecy claimed in the name of *political sovereignty*. Within that is the specific biological warfare problem of "dual use," the possibility that military activities can be disguised as *commercial research*, *development*, and volume production. In the early debates about how to ban *chemical weapons*, this problem was also central. The commercial production of chemical compounds for industry required laboratories and plants that could be converted to military use, for example, German dye works in World War I. Some of the most notorious chemicals (such as chlorine and phosgene) also had peaceable industrial purposes. With *biological weapons*, dual use often refers to large laboratories and pharmaceutical plants, where either legal or prohibited agents might be prepared and grown in bulk. In contrast to chemicals, the agents historically developed for biological weapons have had little or no commercial value that would make it necessary to produce and distribute them in great quantity. Many are also easily found in nature. Even the detection of traces of a dangerous agent such as anthrax or tularemia at a site can raise suspicions, which might be founded or unfounded.

The microbiologist's skills have also been included as part of the problem of dual use. The laboratory scientists who developed biological weapons were generally trained for academic, public health, or commercial employment. What distinguished them was their employment in state programs. A state-sponsored program may not be, after all, easy to hide. It requires a considerable physical plant for handling and storing dangerous pathogens; verifying their virulence in animals; culturing and concentrating them in quantity; safely disposing of wastes; and developing and testing munitions or other devices in chambers. It also requires outdoor space for field tests of munitions and simulations of attacks.

Further, a biological warfare program generates a paper trail. It relies on trade transactions, perhaps international ones; official records; research documentation; the recruitment and training of personnel; organizational integration into military units; and a variety of other activities. If successfully halted, any one or several of these functions could stop a program cold. The secrecy and disinformation that obtained within the totalitarian Soviet Union, though, combined with its resistance to outside contacts and its enormous size, successfully obscured signs of its large program. Cold War intelligence on Soviet biological weapons was necessarily faulty and uncertain. As the USSR program grew, relatively few people even there knew how the risks of proliferation were multiplying.

Origins of the Soviet Program

The historical overview of the Soviet biological weapons program has generally been pieced together from multiple sources, primarily those from German sources documented following World War II, from former Soviet *biological weapons* scientists as the USSR collapsed, from scholars working with newly available archives, and also from recent East-West scientific exchange and demilitarization programs. This information allows a composite sketch of both the old biological weapons program from early in the century and of the later Soviet offensive initiative, called *Biopreparat*, that began soon after the 1972 Biological Weapons Convention was signed in Washington, London, and Moscow.

In 1945, Allied forces captured two Germans who had access to intelligence information about the Soviet program. Col. Walter Hirsch, an Austrian chemist in the German chemical weapons program, wrote a long report for US military intelligence on the USSR chemical program that also includes in formation on the biological weapons program. Another German official, Dr. Heinrich Kliewe, from the German biological program, was also debriefed by US Army intelligence on what he knew about the USSR program.

In late 1989, Soviet program scientist Vladimir Pasechnik defected to the United Kingdom. Although he left no comprehensive overview, information he relayed about Biopreparat, especially research at his Leningrad institute, was the first intimation of the scope and offensive intent of the Soviet venture. In 1992, the former deputy director of Biopreparat, physician Ken Alibek, defected from Russia to the United States; his debriefing and, later, his 1999 memoir of his military career added much contemporary detail. Also in the aftermath of the *Cold War*, retired microbiologist Igor Domaradskij provided information on the program from the perspective of a high-ranking civilian scientist.

In the period following World War I, the Soviet Union carried out and published extensive research on bacteriology in the institutes of the People's Health Commissariat created to stave off recurrent epidemics that were ravaging the country. Russian scientists were deeply influenced by French and German microbiology of the previous century and, relying on research institutes throughout the nation, made important contributions of their own to the control and prevention of infectious diseases such as anthrax, plague, and cholera.

Like other industrialized military powers, the Soviet Union organized its initial biological weapons program on the basis of its chemical program. After World War I, when both Germany and the USSR were politically isolated from Western Europe, they shared air force training and the testing of chemical weapons, ostensibly for defensive purposes. One of the participants in these joint exercises was Jacov Fishman, in 1925 the head of the Soviet *Military Chemical Agency* (MCA).

In 1928, Fishman, who is credited with being the father of Soviet biological weapons, proposed the basic framework that gave the MCA offensive responsibilities and delegated defensive efforts to the institute of Chemical Defense and the Ministry of Health. Fishman's innovations were part of the modernization of the Soviet military which, under the leadership of Marshal Mikhail Tukhachevsky, fostered central command and technical innovation based on Western models.

The joint Soviet-German chemical and air force exercises ceased after Hitler came to power in 1933. The much-publicized 1934 Wickham Steed report of German experiments with biological agent simulants in Paris alarmed the Soviets as well as the British and French. The Soviets also feared that the technologically advanced British might resort to bacteriological weapons, which its officials derided as "a novel type of warfare prepared in bourgeois countries."

Col. Hirsch's principal informant was a captured Soviet scientist who, starting in 1933, had worked at the Scientific Medical Institute of the Red Army in Moscow. By this prisoner's account and others, the Soviet biological weapons program in the late 1930s moved aggressively to establish projects in multiple research institutes near Moscow as well as in the Urals, where during the war military industries were being built beyond the reach of German troops. For example, in Chkalov (Orenburg), a bacteriological research institute appears to have been concerned with the dissemination of bacteria in air and in 1943 another bacteriological facility was built in Sverdlovsk (Ekaterinburg), for military purposes.

Fishman early on proposed that open-air field trials for biological weapons be conducted in remote locations within the USSR. In 1936, Ivan Velikonov, director of the Scientific Medical Institute, which was under Fishman's jurisdiction, led the first of such trials at Vozrozhdeniye (Rebirth) Island in the Aral Sea. Complete with two ships and two aircraft, the trials apparently lasted from spring 1936 to fall 1937.

In 1937, both Fishman and Velikonov were arrested and exiled in Stalin's purge of the military, which extended to public health and medical officials. Many biologists, including the heads of institutes, were arrested on charges of sabotaging the health of the state, and in show trials they confessed to committing sabotage for Japan or Germany by spreading hog cholera, causing anthrax deaths among cavalry horses, and poisoning military food supplies. Also in 1937, Marshall Tukhachevsky, an avid proponent of combined chemical and biological capabilities, was executed for treason. Soviet biological warfare activities after the 1937 purges are less well documented than the declining state of the state's biological science research. During the war, Stalin began supporting Trofim Lysenko, the agronomist who theorized that the environment could change hereditary traits and denounced modern genetics. Lysenko's influence, like that of racialist scientists in Nazi Germany, spelled a significant setback for scientific research and it lasted for decades.

The Soviet's foremost military advocate of biological weapons was Gen. Yefim Smirnov, head of military medical services during World War II and, after the war, the Soviet Minister of Health. Known as "an impassioned advocate of biological weapons" in 1973 he took charge of the Fifteenth Directorate of the Ministry of Defense, the organization that was intended as Biopreparat's "customer" for these innovative weapons. Smirnov retired from this position in 1981 but remained a figure to be reckoned with until his death in 1989.

Yuriy Ovchinnikov, an accomplished biochemist of the next generation and later vice president of the Soviet Academy of Sciences, is often described as the scientific visionary behind Biopreparat. In the early 1970s, he reportedly convinced Soviet Secretary General Leonid Brezhnev and other authorities that the USSR seriously lagged behind Western genetics and molecular biology and should begin a national research program to close the gap. Ovchinnikov may have intended only scientific advances that had pharmaceutical value, without any exploitation by the military.

The totalitarian government of the USSR made this impossible. In 1973, the Soviet Union created Biopreparat as a conglomerate for commercial research and a mask for its secret biological weapons program. Under the Council of Ministers, the USSR's highest governing body, Biopreparat involved the Ministries of Health, Agriculture, and Chemical Industry, as well as the USSR Academy of Sciences and the Academy of Medical Sciences and the KGB. This biological weapons program relied on a dozen research and production centers throughout the country. It employed some nine thousand scientists and technicians, although only the upper echelons fully understood the offensive purpose of its mission. The size of a small city, Obolensk, a closed biological Weapons compound outside Moscow, was built in the 1980s along models developed years before for nuclear-weapons development and production. Post—Cold War revelations about the Soviet program's capabilities, for instance, the genetic manipulation of pathogens and the capacity to produce hundreds of kilograms of anthrax spores monthly, created great concern in the early 1990s.

Biopreparat's top officials and many of its employees were drawn from the Army's Fifteenth Directorate, which exercised great authority in influencing and expanding the Soviet biomedical infrastructure. Biopreparat's Interagency Scientific and Technical Council, which assembled representatives from other bureaucracies, including the KGB, was its "*brain center*," where decisions were made on projects.

As in other biological weapons programs, Biopreparat's main functions were laboratory research to guarantee pathogen virulence and large-scale production for munitions. Using older Western techniques, Soviet scientists experimented with bacteria resistant to antibiotics, antigen-altered bacteria and viruses, and enhanced survival of microorganisms in the environment, for example, improved resistance to sunlight and moisture. Soviet scientists also developed antibiotic-resistant strains of plague and tularemia and attempted to combine pathogens genetically to increase their virulence.

The system supported scientific institutes in Moscow and Leningrad. Other facilities were located at Kirov, five hundred miles east of Moscow, and in Siberia, Uzbekistan, and Estonia. Biopreparat created new *biological weapons* enclaves, such as the one at Obolensk and another at Koltsovo in Siberia (known now as Vector).

Biopreparat built factories for biological-agent production, most impressively a large plant at Stepnogorsk in Kazakhstan, and six other facilities that reportedly could quickly produce large amounts of agents if mobilized. For open air testing, the USSR military relied on Vozrozhdeniye island and other remote sites, just as Jacov Fishman had advised, and maintained four facilities for development and production, including the one at Sverdlovsk. The extent to which biological weapons were assimilated by the Soviet military remains an open question. According to Ken Alibek, if word came from Moscow, the entire system stood ready to produce pathogens, load them into munitions, and launch them abroad. But whether high-level Soviet military commanders actually had war plans incorporating biological weapons, for example, for surreptitious use against Afghanistan or in a last-ditch attack in a nuclear war with the West, is unknown.

The motivation for the Soviet violation of the BWC may never be fully understood. Leonid Brezhnev and civil and military leaders who knew about the program might have perceived a military advantage in developing a type of weapon in which other powers had no interest, perhaps to be used in combination with *nuclear*, *chemical*, or other *weapons*. Or the Soviet initiation and expansion of the program may have been a bureaucratic response to the *Cold War* arms race, a flaunting of the 1972 BWC but completely in keeping with the destructive power of nuclear weapons. By some insider accounts, the Soviets never believed that the United States terminated its offensive program in 1969. Therefore, it sought to keep pace with an American weapons program it suspected was drawing on the major science breakthroughs reported in Western journals. During the 1980s, US defensive research projects were known to involve genetics and cloning and to include a long list of exotic pathogens (including *Ebola virus*, *Marburg virus*, and *Rocky Mountain spotted fever*) and exotic toxins that posed minimal public health threats but could be lethal biological weapons.

After 1975, American suspicions about a Soviet offensive program waxed and waned. In 1978, when a Bulgarian exile, Georgy Markov, was assassinated in London by a passerby using the toxin ricin, the US government suspected that Soviet intelligence had technically assisted in this assassination. The suspicion was correct. But was the Soviet Union capable of concerted use? And how could that be determined?

"Yellow Rain" Allegation

President Jimmy Carter entered office in 1976 with hopes for bilateral arms-control negotiations. Those hopes disappeared in December 1979, when the Soviet army invaded Afghanistan to reestablish a pro-Soviet government and found itself mired in a long and unpopular war. During Carter's administration, rumors began circulating from Southeast Asia about a mysterious and deadly "*yellow rain*" affecting mountain Hmong tribes, allies of the United States during the Vietnam War who had run afoul of the Laotian government. In 1981, after Ronald Reagan became president, the administration began capitalizing on the yellow rain rumors to bolster its confrontational approach to Cold War politics. High on its agenda was support for the army's renewed production of chemical weapons.

In September 1981, in a speech in West Berlin, Reagan's Secretary of State Alexander Haig accused the Soviet Union of abetting the use of lethal trichothecene mycotoxins in Southeast Asia. This accusation was the beginning of a Cold War controversy that resembled the 1952 Korean War allegations. But now it was the United States that was making the accusation, piecing together its case from intelligence and other sources. Also unlike the Korean case, the circumstances of this controversy allowed significantly higher standards for independent expert inquiry. Laos, Vietnam, and Cambodia were inaccessible, and Afghanistan, which figured in a minor way in the accusations, was in the midst of a war with the Soviet Union.

Access to alleged witnesses was nonetheless possible. Hmong refugees had been fleeing Laos for years, and many were living in refugee camps in Thailand. These refugees were in fact the primary source of accounts that Vietnamese or Laotian planes had sprayed their villages and fleeing refugees with a lethal substance that caused sickness and death. When asked for samples of this material, the Hmong provided vegetation and stones with yellow spots on them. Eventually, a university laboratory under US government contract conducted chemical analyses of these and other samples and reported mycotoxins (toxins produced by fungi) as the yellow rain agent. The physical samples, the refugee interviews, and the US government's chemical analyses were each available for independent expert review.

Standards of Evidence

Consulted by the CIA in 1980, Harvard biologist Matthew Meselson was intrigued by the variety and ambiguity of the yellow rain evidence. In 1983 he organized a conference of academic and government experts in Cambridge to discuss the evidence thus far accumulated—about the nature of the toxin agent, where it was found and reported, what scientists had tested it and what witnesses had described the airplane attacks. The first part of the US accusation to unravel was the discovery that the yellow spots on leaves and stones were, in fact, the innocuous feces of Asian honeybees.

The US government had itself conducted chemical analyses of the physical samples but neglected to examine the material under a microscope and to make comparative analyses. Using an electron microscope, botanist and pollen expert Joan Nowicke at the Smithsonian Institution examined yellow rain samples brought to her by the Army and samples sent to her by Meselson, which he had obtained from the Army, the Canadian Department of Agriculture, and ABC News. Her results showed that these samples were largely composed of pollen identical to that in bee feces collected in Southeast Asia in the same ecological environment as the alleged attacks. Onsite in northern Thailand in March 1984, Meselson and bee experts Thomas Seeley and Pongtep Akratanakul documented the common but often ignored phenomenon of yellow "*showers*" from thousands of simultaneously defecating bees. Comparison of these field samples with yellow rain samples showed them to be virtually identical.

Scrutiny then turned to the Hmong interviews. This testimony turned out to have been informally conducted in the refugee camps by untrained medical staff and US military personnel. Interviews under the auspices of the United Nations and the Canadian government added to the sum total of accounts, which, when analyzed, showed great inconsistency in descriptions, often second hand, of the alleged airplane and rocket attacks and their medical effects on victims.

In 1983, the US Department of State and the Department of Defense authorized a joint team to crosscheck Hmong interviews in the refugee camps. During its two years in the field, the team was unable to corroborate attack accounts. Documents later declassified showed that the team ultimately ascribed the principal attack scenario to "*insects* or some other *natural phenomena*."

Secretary Haig's September 1981 Berlin speech had relied on evidence for trichothecene mycotoxins from a single uncorroborated analysis of a leaf and stem sample. Months before the speech, that sample had come not from Thailand or Laos but from Cambodia, accompanied by a muddled story about soldiers falling ill from polluted drinking water. This sample was sent to the Army Chemical Systems Laboratory, where it was divided, with one part sent to toxicologist Sharon Watson of the US Medical Intelligence and Information Agency, housed at Fort Detrick. Watson then sent her sample to plant pathologist Chester Mirocha at the University of Minnesota.

On August 17, in a secret message to the intelligence community, Watson announced that Mirocha had found trichothecene mycotoxins in the material, a type of toxin not on any biological weapons agent list, but she wanted scientific corroboration. On August 31, Richard Burt, then director of the State Department's Bureau of Politico-Military Affairs, instructed a reluctant Watson to prepare a public statement. Her memo on Burt's phone call noted, "Despite these concerns [about corroboration], it was decided to release the information since we were told that it would break anyway with or without our permission."

Mirocha later reported more mycotoxins on other samples, but few questions were asked about his test methods, particularly about controls and false positives, or about the background presence of naturally occurring mycotoxins in Southeast Asia. Nor was any account made public of the results, positive or negative, of the much larger number of additional alleged attack samples and controls he tested.

Following Haig's speech, Britain, France, Canada, Sweden, and the US Army itself commenced a series of careful tests on yellow rain samples. The Army and other government laboratories relied on a powerful and elaborate technique combining gas chromatography and high-resolution mass spectrometry (GC/MS), at a level far more stringent than that employed either in Minnesota or at Rutgers University in New Jersey, where an ABC News—acquired sample had been reported positive. No other methods at the time were likely to be reliable and, without rigorous precautions, even GC/MS could easily yield false positives.

Mirocha's results at Minnesota were not replicated; entirely negative findings from the US Army and from Porton in the United Kingdom that undermined the US case were kept secret for years. Negative results from Porton Down, for example, were withheld for four years. Some US Army data remain secret, which suggests that they may not support the US accusation. Meanwhile, when stringent methods were used, laboratories in Canada, France, and Sweden found no positive evidence for an association of trichothecene mycotoxins with yellow rain or with the blood and urine samples from alleged victims.

Mycotoxins, banned under the BWC, were sufficiently ambiguous to allow the Reagan administration to conflate them with chemicals, so that the yellow rain allegations became background noise for support of renewed US production of chemical weapons. The Army wanted to produce binary artillery projectiles and a binary glide bomb called "*Big Eye*."

These projects were resisted by the US chemical industry, which, after Vietnam, wanted no association of its products with warfare and environmental destruction. Chemical weapons were then being used by Iraq in the Iran-Iraq war, to international criticism; the Reagan administration, with its support of Iraq and the Chemical Corps, was then in no position to protest. Ultimately, the Senate was evenly split in its votes for production of binary projectiles and for the *Big Eye bomb*. In both instances, Vice President George H. W. Bush—later, as president, an advocate of the 1993 Chemical Weapons Convention—cast the tie-breaking vote in favour. With the Soviet Union signaling that it wanted to end the arms race, Congress declined to fund these chemical projects.

Sverdlovsk Anthrax Allegations

In all biological weapons programs, disease risks posed a hazard for workers. Civilians were also at risk for accidental exposure. The Soviet Union experienced two such accidents. In 1972, as the USSR was involved in BWC negotiations, an epidemic of smallpox broke out in a town in Kazakhstan, not far from a biological weapons testing site. Thirty years later, a former public health minister disclosed that the military had caused the outbreak, although the details of the event remain unknown.

In March 1980, learning of a large outbreak of anthrax in the Soviet city of Sverdlovsk, the US government publicly voiced its suspicion that the Soviets were violating the BWC. In response, Soviet officials insisted that the epidemic was caused by anthrax-contaminated bone meal in livestock feed and the consumption of meat from infected animals. They pointed out that anthrax epizootics (outbreaks among animals) were common in the Sverdlovsk region, which is in the southern Urals, near Siberia.

Matthew Meselson was again consulted by the US government and, along with other experts, attempted to evaluate the ambiguous intelligence information about the anthrax outbreak. He found the Soviet explanation plausible but unproven and sought to conduct an onsite investigation. In 1992, after the demise of the Soviet Union, he was able to lead a team of American and Russian scientists to the scene of the outbreak in Ekaterinburg, the former Sverdlovsk. There the search began. The team sought epidemiological evidence through interviews with the families of victims, eyewitness accounts by medical personnel, and hospital records.

Fortunately, the two pathologists who conducted most of the autopsies in 1979, Dr. Faina Abramova and Dr. Lev Grinberg, had preserved tissue samples and case notes from confiscation by the KGB, and this material proved essential in showing that victims died of inhalational not gastrointestinal anthrax. Even more conclusive, epidemiological maps showed that on April 2, 1979, an aerosol emission from a local military facility, called Compound 19, caused around seventy deaths and at least a dozen illnesses among the approximately five thousand people in the area who were exposed.

The team's speculation was that the aerosol release was accidental, due to faulty air filters. The Russian military claimed it had no records of the incident, but Ken Alibek in his memoirs recalled rumors of the military's careless failure to replace an air filter at the facility. The gathering and analysis of information about the Sverdlovsk outbreak, which took nearly two years, showed that a controversial outbreak could be scientifically traced to its source—provided experts were given access and were relatively free from political pressures.

The 1992 Sverdlovsk inquiry also yielded at least two important medical findings about inhalational *anthrax*, a rare disease not usually well documented. One finding was that its incubation period for some victims might be longer than previously thought. It was usually predicted that infection by an anthrax aerosol led to symptoms within two to five days, followed by nearly certain death within two days after that. (In the Japanese experiments at Unit 100, Chinese prisoners had become sick and died even more quickly.)

In Sverdlovsk, the fact that a small number of people became symptomatic weeks after exposure suggested another model, namely, that the anthrax spore in some cases can remain dormant in the human lung for some time, as long perhaps as six weeks, after which it might germinate and cause a fatal infection. During the Sverdlovsk outbreak, once symptoms began (whether two days or six weeks after exposure), death usually did occur two or three days later. Prompt administration of antibiotics could prevent death; Soviet physicians reckoned that they saved fifteen patients this way in 1979, although missing medical records make full case details impossible to confirm.

The 1992 investigation also found that older people were more vulnerable to the disease than younger ones. Many children and young people were in the contaminated zone, yet no one under the age of twenty-four died or appears to have even contracted inhalational anthrax. The older ages of those who contracted inhalational anthrax during the 2001 anthrax attacks supported this finding. The highly unusual Sverdlovsk epidemic is frequently cited as a model for how a covert terrorist attack on an urban population might unfold, eliciting an emergency public health response, from the screening of possible cases to intensive care to the distribution of antibiotics.

Using a *live spore vaccine* called *STI* (different from the US cell-free vaccine), public health officials from Moscow also planned to vaccinate fifty thousand people in the neighbourhood near the military facility. Many residents grew wary when they witnessed the vaccine's side effects, which can include a large ulcer at the point of vaccination and flulike symptoms lasting several days. Around twenty thousand people failed to present themselves for vaccination.

The worst consequence of secrecy and disinformation in Sverdlovsk was the delay in the diagnosis of cases. City officials, ignorant of the military's anthrax project, were taken unaware by the first sudden, mysterious deaths. A week passed before laboratory tests confirmed the diagnosis of anthrax. By then more than a third of those infected, who might have been effectively treated with antibiotics, had died or were dying. Warnings about infected meat misrepresented the true source of danger, which was the military facility. As many as seventeen residents and workers died before they could be hospitalized.

In May 1992 Russian Federation President Boris Yeltsin stated publicly that the military was responsible for the Sverdlovsk anthrax epidemic and decreed that the families of victims should be compensated, a promise left unfulfilled. Further, the language of his decree made it impossible for the military to be held legally accountable for the anthrax deaths. No civil or military hearing was held and, within the Russian military and government, disinformation and misinformation about the cause of the 1979 Sverdlovsk outbreak persisted.

Confidence-Building

In 1990, following Vladimir Pasechnik's revelations about the Soviet biological weapons program, the US and the UK exerted pressure on President Mikhail Gorbachev to admit the USSR had an offensive program. Gorbachev responded with an invitation for them to pick four sites that they would be free to inspect. Intent on verifying Pasechnik's account, they chose four research institutes he had described rather than the controversial Compound 19 in Sverdlovsk. In exchange, a Soviet delegation visited Fort Detrick, Dugway Proving Ground, the Pine Bluff Arsenal, and the Salk Institute in Clearwater, Pennsylvania.

These and subsequent exchange visits went on until 1995, when communication broke down, mostly due to military resistance on the Russian side and resistance from pharmaceutical companies on the US side. Nonetheless, optimism prevailed among parties to the Biological Weapons Convention that Russia would now follow voluntary treaty compliance measures agreed to and modified at the BWC review conferences held after 1975. These agreed measures eventually required six types of declarations, including of all high containment facilities, of defensive research and development programs, of unusual disease outbreaks, of legislation relevant to the treaty, of past activities in offensive and defensive biological research and development programs, and of human vaccine production facilities.

Each declaration has a standard form, with a letter for each category. Confidence-building Form F related to past activities and it was this declaration that was most anticipated from Russia. Early in 1992 President Boris Yeltsin denied any continuation of the Soviet program. In sum, the Fifteenth directorate had been abolished, and

Biopreparat was to become a commercial pharmaceutical venture specializing in diagnostic technology.

In July 1992 the Yeltsin government submitted Russia's confidence-building Form F to other parties to the convention via the United Nations. This document briefly described Soviet and Russian offensive and defensive efforts from 1946 to March 1992. The declaration recounted that at the end of the 1940s the USSR had a defensive biological weapons program. Later, in the 1950s, facilities at Sverdlvosk and Zagorsk were constructed for large-scale agent production. These two facilities, plus one at Kirov, were described as places where work was done on anthrax, tularemia, brucellosis, typhus, plague, and Q fever.

Another facility, the Scientific Research Institute of Military Medicine in Leningrad, was also identified as an important research center. The declaration stated that stockpiles of biological agents were never "prepared or stored" at these facilities, but it affirmed that the plants had been built for mass production of biological warfare agents. Obviously without "justification for prophylactic, protective or other peaceful purposes," in the words of the BWC, the facilities for mass production of biological warfare agents were a gross violation. By inference, biological agents had been tested for virulence, munitions had been designed and tested, and, again, by inference, a doctrine for use had been created.

According to the declaration, no stockpiles were ever created and "accordingly there never has been and is not now any problem of their destruction." At Vozrozhdeniye Island, the document affirmed, tests were conducted on experimental agents installed in mockups of airborne and rocket-propelled missiles and spraying devices. Tests were also conducted on warning devices and technologies for identifying infectious diseases.

Form F described 1986 as the year when the USSR began dismantling its offensive facilities, with Biopreparat transferred from the Council of Ministers to the Ministry of Medical and Microbiological Industries. By April 1992, the testing facility at Vozrozhdeniye Island was closed, the conversion of facilities to peacetime uses began, and the biological weapons program was declared terminated. Concerning toxins, Form F noted that the program had investigated botulinum toxin and then ended with this terse conclusion, certainly in reference to the US yellow rain allegation: "In the opinion of experts, mycotoxins have no military significance."

Controlling Scientists and Pathogens

In 1988 and thereafter, as the United States and the Soviet Union reached agreements on nuclear arms control and the Cold War ended, the problem of Soviet compliance with weapons treaties receded. Accords between the United States and Russia continued to emerge and public anxiety about a nuclear holocaust, which had risen during the Reagan administration, subsided. The threat of major powers ever again using chemical weapons also abated. In 1993 in Paris, with support from President George H. W. Bush, more than a hundred nations signed the *Chemical Weapons Convention* (CWC). This treaty came with compliance measures that the Biological Weapons Convention lacks: an organization based in the Hague, teams of experts, and provisions for mandatory declarations, inspection, and stockpile destruction.

With the Cold War ended, and with Yeltsin's stated cooperation, the former tension around Russian treaty compliance was much reduced. Dismantling the Soviet biological weapons program remained a complex problem. Questions were raised as to whether Gorbachev or Yeltsin had full authority over powerful Biopreparat directors and their military counterparts. During the 1990s, enterprising plans to convert old Soviet biological weapons facilities to pharmaceutical production sprang up and then foundered. A decade after Yeltsin's declaration that Russia had no biological weapons program, the three major military facilities at Yekaterinburg, Sergiev Posad, and Vyatka remained closed to outside observers. The virology center at Sergiev Posad had been the largest and most secret installation and was reported to have experimented on the smallpox virus.

Ensuring openness so that Russia would not revert to an offensive biological weapons program was another main problem. A third fear was that former Soviet biological warfare scientists might be recruited to programs in lesser states hostile to American interests. In 1991 Senator Sam Nunn and Senator Richard Lugar took legislative initiative with the Nunn-Lugar Cooperative Threat Reduction Act. This bill funded the conversion and destruction of former nuclear, chemical, and biological facilities and joint projects with Russian former weapons scientists. The US National Academy of Sciences helped to select and guide worthwhile collaborative projects biology; the Centers for Disease Control, the Department of Energy, and the National Institutes of Health joined in the effort. Another perceived threat from the Soviet program was that dangerous pathogens would proliferate. A strain of anthrax developed by Russian scientists proved in animal tests to be resistant to the standard Russian vaccine, although not to a modified vaccine. Another dangerous pathogen was smallpox. Following the 1980 eradication of smallpox by the World Health Organization (WHO), only the Soviet Union and the United States were permitted to have the smallpox virus; the former kept its strains at an institute in Moscow, the latter at the Centers for Disease Control in Atlanta.

In the 1980s, the Soviet Union had secretly moved its smallpox reserve to Vektor in Siberia, ostensibly to guard it against terrorists. Failure to inform WHO of this change at the time rankled some in the West. According to Alibek, whose account was disputed, as late as 1990 Vektor scientists had been developing genetically altered strains and had the capacity to produce vats of the virus. The institute preserved fifteen thousand isolates of pathogenic viruses, including some hundred and twenty strains of smallpox, fifty of them unique to Russia. The possibility that loose security would allow unauthorized persons to steal or buy one of the Russian strains was worrisome, especially since smallpox vaccinations had long ago ceased throughout the world.

Some critics felt that security at Vektor was inadequate and that the facility remained ominously secret. Still others felt that joint projects with the West were the best way to reduce the risks of proliferation. Starting in 1999, Vektor and the CDC together embarked on a series of smallpox research projects endorsed and supervised by the WHO and funded at over $4 million by the United States Department of *Health and Human Services* (HHS) and the *Defense Threat Reduction Agency* (DTRA) within the Department of Defense.

With reforms, collaborative projects, and the gradual replacement of Soviet-era figures with younger political actors, the biological warfare security risks associated with Russia were diminishing, although not with ideal consistency. Suspicions of Russian motives and handling of US funds for threat reduction persisted, and there remained much to learn about cooperative program.

Still, an important change had taken place: biological weapons were finally uncoupled from the strategic programs of the two major Cold War adversaries and, indeed, from all major state militaries. The central problem became protecting civilians from biological weapons proliferation from any source and especially from hostile developing nations and terrorists. A reinforced Biological Weapons Convention was first on the list of potential restraints. In 1991 Western European nations took the lead in proposing mandatory compliance measures, analogous to those being proposed for the new Chemical Weapons Convention. The United States, protective of its military and industrial secrecy, balked at required and other measures promoting transparency. In 1994 an Ad Hoc Group was created to formulate a legally binding protocol for the BWC that would overcome these objections by the United States and other countries. The process would prove difficult. The question was whether the United States could balance its new role as the lone superpower in international affairs with its national defense policies.

5

SECURITY AND THREAT

The September 11, 2001, Al Qaeda attacks on the United States and the anthrax postal attacks that soon followed together intensified Clinton-era policies already in place for national security and defense against biological weapons. The federal intelligence budget rose again. The domestic preparedness programs in place throughout the United States received more support, as did funding for technological defenses, from air sensors to new pharmaceuticals. The great change after 9/11 was in the organizational scale of the Bush administration responses to this unprecedented foreign attack on Americans. The minor use of force, that is, the 1997 bombings of Afghanistan and Sudan, was overshadowed by military invasion in the name of the war against terror, whether the threat was training camps or Iraq's weapons of mass destruction.

The 2003 creation of a Department of Homeland Security far outweighed the diffuse, decentralized domestic preparedness project of the previous decade. Clinton's fortification of a few public health technologies was dwarfed by sweeping policies to incorporate US biological sciences into the campaign against bioterrorism. Each of these expanded policy initiatives brought dilemmas concerning the flow of information and public trust in government. The invasion of Iraq was eventually followed by disclosures suggesting that the Bush administration manipulated intelligence data in order to convince Americans of the imminent threat of Saddam Hussein.

The violent and costly military occupation that ensued further troubled public trust in the administration. The Department of Homeland Security from the very start faced the problem of relying on intelligence information to pronounce public alerts that were credible, even when no crisis occurred. Could a state of emergency be indefinitely maintained? In addition, there was the question of whether local first responders would be in the Washington information loop or, as happened during the anthrax letter attacks, would physicians and targeted victims be the last to know about the risks? Last, the large-scale government recruitment of biologists to the biodefense campaign signaled new rules for secrecy in basic scientific research that, until the present, has been open in its beneficent intents and goals.

Security and Defensive Research

Even before September 11, the presidency of George W. Bush signaled a return to Reagan-era policies, which emphasized use of force and technology for national defense, less as a complement to international agreements than as a substitute for them. In charge of the world's mightiest military nation, the Bush administration proved much more disdainful of arms-control treaties than President Reagan or the former President Bush. In July 2001, it withdrew from negotiations for a legally binding verification and compliance Protocol to the Biological Weapons Convention. It stood opposed to ratification of the Comprehensive Nuclear Test Ban Treaty, and in July 2002, it withdrew the United States from the 1972 Anti-Ballistic Missile Treaty. These decisions and the 2003 unprovoked invasion of Iraq contributed to the perception abroad that the United States claimed exception from international norms against armed aggression.

From the beginning, the Bush administration took the position that military secrecy and pharmaceutical commercial interests took precedence over negotiations to strengthen the *Biological Weapons Convention* (BWC) with mandatory compliance measures. In December 2001, having rejected negotiations, the Bush administration put pressure on the Ad Hoc Committee deliberating the protocol to disband entirely. In a compromise, Hungarian ambassador Tibor Toth, chair of the committee, agreed to a three-year plan for continued discussions on implementing the treaty, with verification off the agenda. The issues for discussion became limited to subjects that did not trouble military or commercial interests and that states could pursue independently of a general accord.

These agenda items were national enactment of penal legislation; national oversight of dangerous pathogens; measures for investigating and responding to allegations and suspicious disease outbreaks; the surveillance and combating of infectious diseases of humans, animals, and plants; and the development of codes of conduct for scientists.

Washington's resistance to transparency was better understood when, in September 2001, the book *Germs*, written by three investigative journalists, revealed secret US projects that blurred the line between defensive and *offensive development*. These projects included the making and testing of a model of a Soviet anthrax bomb, the building of a small biological-weapons production facility from material bought on the open market, and the creation of a vaccine-resistant strain of anthrax. They were kept so secret that the first two projects were unknown to the Clinton National Security Council staff and to the president himself until they were tipped off by the journalists, who at the time had yet to reveal the information to the public.

For the bomb project, the CIA consulted with scientists, among them Joshua Lederberg, who suggested a legal review regarding possible violation of the BWC prohibitions. CIA lawyers decided the project was within the allowed realm of defensive research. CIA lawyers decided the same for the second project, on the small, fully functional biological weapons facility, sponsored by the Defense Threat Reduction Agency (DTRA) within the Defense Department. The third project, developing a vaccine-resistant strain of anthrax, was meant to duplicate a Russian experiment done in the early 1990s and published in the open literature.

When the CIA proposed that the United States make its own version, the Clinton administration hesitated. In the first months of the Bush administration, National Security Council officials gave their permission. They believed the Pentagon had the right to investigate genetically altered pathogens in the name of biodefense, "to protect American lives." The Pentagon then authorized the *Defense Intelligence Agency* (DIA), under its secret "*Project Jefferson*" program for pathogen re search, to re-create the strain. The news of these projects confirmed the administration's exceptionalist national security position. They showed that the United States military and intelligence agencies had the power to stretch the limits of the treaty without public review and beyond what would be tolerated for other countries.

Iraq and Weapons of Mass Destruction

After September 11, the members of the UN Security Council and the international community generally supported the US invasion of Afghanistan later that year and the destruction of the Taliban government. All of the major powers had experienced terrorist violence against civilians and their sympathy for American losses from the Al Qaeda attacks ran high. The Security Council, however, became deeply divided over the US planned invasion of Iraq in 2003. For some, this aggression indicated a dangerous disregard for international legal restraints. But no nation or alliance of nations was in a position to countermand the US decision and the United Kingdom sided with the Bush administration. At the highest level of US policy making, the decision had already been made to topple the government of Saddam Hussein.

In their appeals for public support for an invasion, President Bush and British Prime Minister Tony Blair portrayed the imminent danger of Saddam's weapons of mass destruction as the primary justification for war. In coordination with a speech of Blair's to Parliament, his office released an intelligence summary to the London Evening Standard, which, on September 24, 2002, headlined that Saddam was "45 minutes from attack" with chemical weapons capable of reaching British military bases in Cyprus. Meanwhile, France, Russia, and Germany were arguing for continued containment of Iraq and more time for UN inspectors to complete their mission. To forestall an invasion, on November 8, 2002, the UN Security Council adopted Resolution 1441, requiring immediate, unconditional, and active cooperation from Iraq. Shortly thereafter, UNMOVIC (United Nations Monitoring, Verification and Inspection Commission), the successor to UNSCOM, began onsite inspections in Iraq.

The Bush administration's resolve to destroy Saddam Hussein's regime may well have been set years before, completely apart from 9/11. In 1991, George W. Bush's Vice President, Richard Cheney, had been Secretary of Defense for President George H. W. Bush. At that time he articulated an aggressive containment policy for dealing with rogue states, whereby they should expect massive military retaliation for their aggression and a bombing of their weapons of mass destruction "out of existence." The cease-fire with Iraq halted the Gulf War before this policy could be carried out, but Cheney had a new opportunity to follow through in 2001.

The series of inspection updates from UNMOVIC failed to persuade the US government that Iraq was no longer an immediate, serious threat. On January 27, 2003, Hans Blix, the executive director of UNMOVIC, reported to the United Nations his evaluation of progress thus far. He said, "Iraq has on the whole cooperated rather well.... The most important point to make is that access has been provided to all sites we have wanted to inspect." UNMOVIC's one hundred trained inspectors had made three hundred inspections, twenty to sites never before visited. It had recently acquired the use of eight helicopters that helped speed inspection. Germany had promised assistance with unmanned surveillance vehicles to monitor changes. Blix also described incidents when Iraqi officials were uncooperative, although they seemed more willing to assist as pressure mounted. About biological weapons, Blix reported that nothing had been found but that a store of anthrax "might still exist" and "should be found and be destroyed under UNMOVIC supervision or else convincing evidence should be produced to show that it was, indeed, destroyed in 1991."

In early February, Secretary of State Cohn Powell made a ninety-minute presentation at the United Nations that, complete with slides and audiotapes, included US intelligence data on Iraq's programs. Special emphasis was put on mobile facilities for producing germ weapons and photographs were shown of trucks, vans, and boxcars allegedly involved in this venture. Holding up a small vial of anthrax simulant, Powell reminded his audience that this same amount, "a few teaspoons' in the 2001 anthrax letters had killed five people and caused great fear and disruption.

Powell failed to say that UNMOVIC investigators had methodically followed through on US intelligence and found no evidence of chemical or biological weapons at suspected sites; nor had they found evidence of mobile facilities. Shortly after, Blix again reported to the UN and raised questions about faulty US and UK intelligence. At the same time, Mohamed Elbaradei, director general of IAEA (the International Atomic Energy Agency), reported that the agency found no evidence of ongoing nuclear or nuclear-related activities in Iraq; his organization had supervised the removal of nuclear materials in 1995 and, under containment policies, no further production had occurred. Powell's assertion that it was a "*fact*" that Iraq had a nuclear program left the erroneous inference that Iraq's nuclear threat was imminent. This assertion was much repeated by officials in the Bush administration.' Still, it remained that the Iraqis under Saddam Hussein had failed to be fully cooperative.

Disregarding the reports of UN inspectors, the United States, along with UK forces, invaded Iraq on March 19, 2003. That May, President Bush declared the cessation of major military operations. For many months after, no weapons of mass destruction or long-range missiles were discovered in Iraq. In January 2004, David Kay, a former IAEA investigator in charge of the US Iraq Survey Group, concluded after six months of searching that Saddam's WMD arsenal had been destroyed in the 1990s, during the UNSCOM investigation."

In the United Kingdom, criticisms of Prime Minister Tony Blair for exaggerating the Iraq WMD threat were severe, and two members of the Blair cabinet resigned in protest over the issue. By June 2003, news headlines were dominated by accusations that Blair had manipulated intelligence information to justify British participation in the invasion. In July 2003, respected former UN inspector David Kelly, a consultant for the UK Ministry of Defense as well as the Foreign Office, committed suicide after being publicly embroiled in the controversy. The national commission, headed by Lord Hutton, that investigated the circumstances of Kelly's death generated even more headlines, although it found no evidence of foul play and avoided blaming Blair's staff for deliberately misinterpreting intelligence data.

Around this same time in the United States, Bush administration's use of intelligence information was only mildly criticized in the press. In his January 2003 State of the Union address, President Bush had claimed that Iraq had sought uranium from an African state. In July, news leaked that the CIA had not authorized that claim; rather, one of the National Security Council staff had inserted it in the speech. Another administration claim, that Iraq had purchased special aluminum tubes for nuclear-weapons production also turned out to be false. The Bush administration's restating of why the war was necessary—because Saddam Hussein was a ruthless dictator who had used weapons of mass destruction and might do so again—was generally accepted as sufficient justification for a war that seemed to have ended victoriously.

Anthrax Letters and Government Response

During the 1990s, the US Government Accounting Office and various critics pointed out that domestic preparedness programs were inexcusably uncoordinated and would be ineffective in the event of WMD terrorist attack. In the autumn of 2001, government responses to the anthrax letters provided and unpredicted real-life demonstration of organizational problems and their repercussions for victims. The anthrax postal attacks blindsided government agencies, which in their responses pursued sharply different goals. The Centers for Disease Control had an epidemiological mission as well as the responsibility to contain the outbreak. The FBI was intent on a criminal investigation. Military and intelligence agencies had roles to play, along with the US Postal Service, the Environmental Protection Agency, and a host of other federal and local responders. The flow of information was frequently blocked, within and across bureaucratic divisions. At the height of the crisis, from mid- to late October, as many as three federal press conferences were held a day, each conveying information that was often confused and confusing.

The anthrax letters were likely mailed in two batches from letterboxes in Princeton, New Jersey. The first set of perhaps five letters was posted around September 18 to mass media offices, one to American Media Incorporated (AMI), a tabloid publisher in Boca Raton, Florida, and the rest to broadcasters and editors in New York City. All these early letters were dismissed as routine hate mail or the letters were left unopened On October 3, 2001, as soon as the first victim, Robert Stevens, was diagnosed in Florida, teams from the Centers for Disease Control in Atlanta were dispatched to investigate the source of Stevens's infection. Their first hypothesis was that he had caught the disease from spores in soil, with little attention paid to the possibility of exposure to a biological agent from any source. When the CDC discovered anthrax spores at Stevens's workplace, the AMI building, the FBI assumed authority and the disease event was framed as a criminal investigation. Evidence emerged that Al Qaeda operatives might have been in the area, not for from AMI. The idea that the source could be a US government-connected scientist only gradually took hold, at least in the public mind.

For the CDC, anthrax was a rarity that posed no public health hazard. The FBI was familiar with limited aspects of anthrax spores as select agents. Since Secretary of Defense Cohen's 1997 speech on anthrax, the agency had been notified of hundreds of hoax letters containing sugar or other powders. The most comprehensive knowledge of anthrax had been developed by the US offensive program, but most of those institutional memories were lost. Before the anthrax letters, the CDC and the FBI were unlikely to share information, although the FBI and Fort Detrick staff did communicate about potential bioterrorist threats. In addition, the postal attacks were such a novelty that only a few officials imagined how mailed spores might be dangerously dispersed.

In 2000, after hoax anthrax letters were mailed to the Canadian Parliament, scientists from the *Defense Research Establishment at Suffield* (DRES) used *B. globigii* (BG) spores in experiments on how anthrax spores might disperse from an envelope opened in an ordinary office setting. The DRES scientists found that even a tenth of a gram of BG dispersed into the air throughout the office space far beyond all previous predictions, and that a person breathing the air would have inhaled spores in numbers thousands of times the estimated human

LD_{50} (the lethal dose for half an exposed population). Such a letter also might, the report suggested, leak spores from unsealed edges. This initially classified information was shared with the FBI and US military but the CDC was unaware of it. On hearing about the first Florida case of anthrax, the DRES scientists emailed the report to the CDC, where officials were too understaffed to read it until the anthrax postal crisis was nearly over.

The DRES report speculated about spores leaking from the unsealed corners of an envelope. No one imagined or tested the possible effect of mechanized postal sorting machines or whether anthrax spores, a micron in diameter, would be able to escape through envelope paper, which commonly has imperceptible openings as large as three microns. Few officials knew about the air turbulence that characterizes postal facilities. Without this knowledge, the calculation of risks to postal workers was greatly impeded.

An even greater hindrance to assessing risks from the anthrax letters was an unnecessary confusion about dose response. Officials at the Department of Health and Human Services and within it, the CDC, had little know! edge of the quest of the old biological weapons programs and their ways of creating anthrax aerosols with spores in the optimal one- to seven-micron range. During the crisis, retired Fort Detrick scientists who had researched this problem in detail were not consulted. They could have told CDC officials that, over the years, the Army routinely used an LD_{50} of eight thousand to ten thousand inhaled spores to estimate standard doses on which to base munitions fill. This extrapolation from animal research was never intended to convey a threshold below which an exposed individual was safe. Although very unlikely, even a single anthrax spore or a small number deposited in the lungs might successfully germinate and cause infection and death; estimates of inhalation per person in the ceramics factory affected by the 1979 Sverdlovsk release, based on US Army methods of calculation, were as low as nine spores per victim. Yet, as if an environment with low levels of anthrax contamination posed no jeopardy, government officials and the media repeated eight thousand to ten thousand spores as a threshold range. And how would a low level of contamination be measured? Neither the CDC nor the FBI was practiced with environmental sampling techniques on which to gauge the risks of anthrax.

A success in the federal response to the anthrax letters was the mobilization of the National Laboratory Response Network. Established in 1999 the network includes one hundred public health laboratories coordinated with the US Army Medical Research Institute of Infectious Diseases (USAMRIID) and CDC. This network absorbed much of the burden of analyzing anthrax hoax materials that flooded local agencies and the FBI by the thousands during the postal attack crisis.

In responding to the anthrax postal attacks, the CDC also showed that the emergency distribution of pharmaceuticals and simple tests for exposure could be accomplished with relative efficiency. Throughout the postal anthrax crisis, CDC and local health officials were able to draw on supplies from the national stockpile established in the Clinton administration. Some 33,000 people received postexposure drugs, usually *Ciprofloxacin* (often referred to as Cipro) or *Doxycycline*. It is difficult to say whether all these people were at risk of exposure, but some were surely protected from infection and possibly death. The problem was in the timing of the distribution of antibiotics, which had to be early, before the anthrax bacteria produced their deadly toxins. And this timing depended on an alert and informed public.

In Washington

It was not until October 12, 2001, a week after Robert Stevens died, that it was certain that letters were the source of multiple anthrax infections. By that time dozens if not hundreds of people had been exposed. On October 12, an anthrax- laden letter addressed to NBC news anchor Tom Brokaw was discovered in New York. After a general search of media offices, involving both CDC and the FBI officials, another letter, this one unsealed, was discovered at the offices of the New York Post. Two cutaneous cases among New York media workers, initially misdiagnosed, were then identified; the infant son of an ABC news producer also contracted severe, misdiagnosed cutaneous anthrax, following a single visit with his mother to the news office in September.

Before the Brokaw letter was found, the criminal who sent the first set of letters may have been frustrated that none had been discovered and reported on by the media. A second set of anthrax letters was mailed around October 8 from Princeton. Two went to Democratic senators in Washington, D.C. One of these was opened on Monday morning, October 15, in the Hart Senate Office Building office of Tom Daschle, then the Senate majority leader. Within the hour, Capitol Hill police had identified the powder as anthrax and staff from the Office of the Attending Physician began distributing Ciprofloxacine to around forty Senate employees at high risk of exposure, with advice to take the pills for sixty days. In the following two days, antibiotics were distributed to hundreds of other Senate and government workers at the Hart Building, which was more contaminated than officials first realized.

The spores (less than a gram) remaining from the Daschle letter were immediately transported to Fort Detrick and analyzed. The next day, in a press conference, Senator Daschle characterized the anthrax letter material as "*weapons-grade*," the description just used in a closed interagency briefing for Congressional leaders. From the White House Homeland Security office, former Pennsylvania governor Tom Ridge, its head, objected to this characterization of the spores; yet it was true that, even in comparison to the spores in the first set of letters, the Daschle letter material was exceptionally free of bacterial residue and dispersed easily.

The technique for its preparation was sophisticated and pointed to a skilled scientist and a state-sponsored program. When the mailed anthrax was identified as the Ames strain, the American program, which had years before named and experimented with it still, was further implicated. More tests by the FBI in 2002 showed that the spores had been prepared within the prior two years of their being sent. At that point, although the FBI appeared to suspect a former Fort Detrick scientist who subsequently worked on intelligence projects, the criminal investigation stalled. The FBI, in developing its profile of a disgruntled American microbiologist as bioterrorist, changed the way that the government looked at this formerly benign category of scientists. Even an American laboratory worker might pose a threat to national security.

Brentwood Postal Employees

The dispersal of anthrax spores put postal workers at serious risk, primarily those where the letters were mechanically processed and coded, at the Hamilton, New Jersey, facility which served Trenton, and at the main Washington, DC, facility called Brentwood after its street address. Of the twenty-two people infected with anthrax, nine were US postal employees. Of the eleven who contracted inhalational anthrax, seven were postal workers. Another victim worked in the State Department mailroom, where the second Washington letter, addressed to Senator Patrick Leahy, had been diverted by mechanical error. Still another victim was an AMI mailroom worker.

In the week after the Daschle letter was opened, evidence of leakage from it and other letters accumulated, but not with the effect of alerting postal workers. On Wednesday, October 17, anthrax spores were found in the mailroom in the adjacent Dirksen Senate Office Building, which served the Hart Building. Senate offices were shut down, as, for cautionary reasons, the House offices had been closed the day before; the Hart building would take five months to decontaminate. In the next two days, there were more warnings. In Florida, a Boca Raton postal worker was diagnosed with cutaneous anthrax. In New Jersey, at the Hamilton facility, nasal swab tests revealed that twelve workers had been exposed to anthrax spores; another worker was diagnosed with cutaneous anthrax. Material from the second set of letters, which seemed to disperse more easily than the first, was the likely source of infection. On Friday, another cutaneous case was diagnosed, which led to the closing of the Hamilton facility and the distribution of antibiotics there.

The US Postal Service, under pressure from Congress to become cost-efficient, had no incentives to close any facility. The 500,000-square-foot Brentwood facility employed nearly two thousand workers and processed much of the federal government's mail. Unaware of the capacity of anthrax spores to disperse and infect at low doses, the CDC officials reckoned that the probability of infection was low. Postal work continued without the distribution of antibiotics or warnings to workers to be alert for symptoms of the disease.

Four of the five inhalational anthrax cases in Washington were Brentwood workers. The first case was that of Leroy Richmond, who was clinically diagnosed on Friday evening, October 19, after being misdiagnosed that afternoon by an HMO physician. This physician, not unlike others during this crisis, made no connection between the patient's workplace, the anthrax crisis, and the patient's flulike symptoms. As the facility that had processed the Daschle letter, Brentwood had been in the news all week. On October 12, within an hour after the Daschle letter was processed through DBCS (Digital Bar Code Sorter) #17, Richmond, who had lent his face mask to another worker, was cleaning the area around the machine when yet another worker, appropriately masked, used air pressure to clean it. Richmond's insistence on a hospital referral brought him to a hospital emergency ward where a physician finally made the connection to Brentwood and alerted the CDC to his case.

On Saturday, October 20, another Brentwood worker was hospitalized and clinically diagnosed with inhalational anthrax. Meanwhile, the CDC waited for laboratory confirmation of Richmond's diagnosis from Atlanta, which was communicated at 7:00 on Sunday morning. That same morning, another floor worker at Brentwood, William Morris, called 911 and died of inhalational anthrax shortly after being admitted to a Washington hospital. Also that morning, a fourth worker, Joseph Curseen, who worked at DBCS #17, collapsed at church. He was diagnosed at a local hospital as suffering from dehydration and sent home. Later he was transported by an emergency ambulance to a hospital where the next day he, too, died of inhalational anthrax.

When the Brentwood postal facility was closed the early afternoon of Sunday, October 21, its distraught employees were aware that three co-workers were hospitalized. Perhaps to maintain calm, the CDC and Washing ton health officials communicated little sense of urgency in the distribution of antibiotics. A rock concert had blocked access to the nearest hospital, so workers were directed to a downtown building where, from a temporary station, they could get antibiotics and have nasal swabs. Or, officials told them, they could wait and go to the hospital the next day.

The failure to communicate the risks of inhalational anthrax to the postal workers and to describe its symptoms cost lives, as did the lack of clear communication with local physicians to be on the watch for any patients from postal facilities. The CDC and others measured the small numbers of those who got sick or died against the larger numbers who, without antibiotics, might have been casualties. One analyst concluded that the public health response was "rapid and comprehensive, and it may have prevented the further spread of anthrax." The underestimation of danger at Brentwood can also be construed as biomedical example of how organizational thinking can obscure and then increase the likelihood of serious risks.

The random deaths of two more people from inhalational anthrax pro longed national anxiety about postal contamination. Sixty-one-year-old Kathy Nguyen, a hospital worker in New York, died in late October 2001, perhaps from contact with cross-contaminated mail. In late November, ninety-four-year-old Ottalie Lundgren died of inhalational anthrax in rural Connecticut. The post office that served her district proved to be contaminated with anthrax, suggesting that trace amounts on her mail were the likely source of her infection. The spores from the Leahy letter contaminated diplomatic pouches, including one sent to Ekaterinburg (formerly Sverdlovsk). There, to verify the source of the spores, the United States relied on some of the same public health officials who had been involved in containing the 1979 outbreak.

In February 2002, former Brentwood employees, redeployed to other facilities, organized a support group, Brentwood Exposed, to deal with the deaths, illness, stress, and reactions to antibiotics caused by the anthrax letters, for which no federal agency took responsibility. That no perpetrator had been found contributed to the community's loss of trust in law enforcement and government. Their workplace took two years to decontaminate, and many workers were reluctant to return.

Judicial Watch, a Washington legal advocacy group started by a former Justice Department lawyer, took up the case of the Brentwood workers. In December 2002, it acquired a Brentwood manager's diary describing his understanding four days before the facility closed that "*hot*" anthrax spores had contaminated it. Judicial Watch then filed a class action suit on behalf of the Brentwood workers and pressed for a criminal investigation into the government's handling of the crisis.

Department of Homeland Security

On November 25, 2003, President Bush signed the Homeland Security Act into law. More than twenty-five agencies were affected, and tens of thousands of government employees would be relocated to the new department. Congress quickly approved Bush's choice for Secretary of Homeland Defense, Tom Ridge.

The stated mission of the new Department of Homeland Security is to "prevent terrorist attacks within the United States, reduce America's vulnerability to terrorism, and to minimize the damage and recover from attacks that do occur." It incorporated the Immigration and Naturalization Service as a way to monitor and protect foreign visitors. It also incorporated FEMA (*Federal Emergency Management Agency*) and assumed responsibilities unrelated to terrorism—for victims of floods, hurricanes, earthquakes and other providential catastrophes, and for industrial accidents.

Since DHS is charged with mediating intelligence data on threats to the nation and conveying threat levels (red, orange, and yellow, for example) to the nation, the department's activities are often protected by secrecy. Reports of many of its activities are unavailable through Freedom of Information Act requests. Its advisory meetings with contractors are exempt from FACA (*Federal Advisory Committee Act*) requirements for public disclosure.

The general trend after September ii was to increase secrecy around federal bureaucracies. The CDC, the nation's public health agency, was for the first time allowed to classify its planning sessions and reports. Counterterrorism activities at the FBI also increased, following a budget increase from $595 million in FY2001 to $1.06 billion in FY2002.

In addition, activity at the Defense Department relative to homeland security expanded. By 2003, the Defense Department had twenty-three different sub-offices and groups for WMD disaster response. As part of the military's increased role, in October 2002 US Northern Command (USNORTHCOM) at Peterson Air Force Base in Colorado was established both to coordinate protection of land, maritime, and aerospace approaches to the United States and, in the event of a WMD attack, to providing civil authorities with "*technical support*" and "assistance to law enforcement; assisting in the restoration of law and order; loaning specialized equipment; [and] assisting in consequence management."

In one effort, the Defense Department also began applying its Advanced Concept Technology Demonstrations (ACTD) for command-and-control military operations to terrorist attack simulations. In December 2002, a complex ACTD exercise simulated twenty related terrorist attacks from New York to Hawaii, with participation by fifty federal agencies, including US marshals, the FBI, the CIA, and the Bureau of Alcohol, Tobacco and Firearms. "The event became the largest real-life, on-the-ground, interagency technical interaction ever conducted for homeland security." The ACTD exercises were later coordinated with DHS as the military's contribution to civil defense, from the federal to the state and local levels.

The DHS maintained the federal commitment to local domestic preparedness, providing funds for emergency personnel, exercises, and equipment and demanding more rational response plans from states. However, due to the economic downturn that began in 2001, the local programs were undercut by layoffs in police, firefighters, and other government workers who routinely safeguard the public. The federal government organization of counter bioterrorism also began raising ethical and legal concerns about individual citizens' rights during epidemic emergencies.

DHS also continued to evaluate federal organizational response by using simulated attack scenarios. In May 2003, in cooperation with the State Department, it sponsored the TOPOFF 2 exercise, which, like the first one in 2001, was to test how federal agencies performed in emergencies. The scale of the fictive attacks increased geographically from the first TOPOFF, which was city-based, to large northern metropolitan areas that extended into Canada. The TOPOFF budget also increased, from $ 10 million to $16 million.

The 2003 exercise simulated a radioactive attack in the Seattle metropolitan area and a related pneumonic plague attack affecting greater Chicago. Nineteen federal agencies, the Red Cross, and the government of Canada were involved in the five-day exercise. The fictional perpetrator of the attacks was an international terrorist organization called GLODO. When it was over, aspects of the exercise were classified, an indication that TOPOFF yielded in formation about national vulnerability that officials deemed should be kept from the nation's enemies and from the American public.

The Smallpox Vaccination Campaign

During the second Bush administration, advocates for domestic preparedness and civilian biodefense continued to talk about a revitalization of American public health. Yet public health continued to mean, even more than in the Clinton administration, a technological approach to national defense. In the Bush administration, pharmaceutical protection became the centerpiece of biodefense policy.

On December 13, 2002, convinced of the Dark Winter—type threat of smallpox, President Bush announced his nationwide smallpox inoculation program. Publicity about Iraq's potential biological arsenal, especially in the lead-up to the 2003 invasion, and the threat of bioterrorism had convinced many in the public to participate. The states and the CDC were ready to handle the logistics. In addition, civilian participation was voluntary, which reduced legal liability for those who administered the vaccine and for the government.

As might have been predicted, this smallpox vaccination campaign found it difficult to circumvent the well-known fears of vaccination as a source of bodily pollution and the mistrust engendered when vaccines appear a worse health risk than the forecast epidemic. The swine flu vaccination program during President Gerald Ford's administration was a controversial miscalculation that killed two dozen and sickened hundreds and, because the flu never struck, it caused widespread distrust in government health initiatives. Reliance on intelligence calculations can be a necessary but problematic basis for predicting epidemics, even for the military. The Pentagon's universal anthrax vaccine program in the 1990 resulted in dishonorable discharge for more than four hundred soldiers who refused to be vaccinated and it caused dozens of National

Guard pilots to resign, out of fear of serious, unpredicted side effects that appeared worse than any biological weapons threat.

The first phase of the smallpox campaign was the mandatory vaccination of 500,000 military and government employees who might be deployed to the Middle East or other potentially high-risk areas. Starting January 2003, Phase One also included the voluntary vaccination of 500,000 "*front-line*" civilian health workers and first responders on specialized "*Smallpox Response Teams*." In the second phase of the smallpox vaccination program, in March-April 2003, ten million additional healthcare workers and first responders could opt to be vaccinated. In the fall of 2003, in Stage Three the vaccine would become available to the American public at large. (Phases Four and Five were emergency strategies for containment in the event of a smallpox outbreak, which meant quarantine and mass vaccination.)

Unlike the anthrax vaccine, the most available US smallpox vaccine, called Dryvax, had a well-known history. Administered with a bifurcated needle, it took six to eight days to be effective. If given within four days of exposure, the vaccine might significantly reduce the chances of sickness and death. The typical reactions to the smallpox vaccine ranged from soreness at the vaccination site to headaches, swelling of the lymph nodes, and fatigue. Brain swelling (encephalitis) was a known but rare reaction. The death rate estimated for universal smallpox vaccination could theoretically be as high as one percent: if ten million people were vaccinated, some ten thousand might die, unless risk factors were recognized in advance. The vaccine was contraindicated for those with eczema or other skin diseases, for pregnant women, and for those whose immune systems were compromised or who were taking immune-suppressing medication.

Identifying these risk factors posed problems. For example, the CDC estimated that around 300,000 Americans were unknowingly HIV-infected. The smallpox vaccination guidelines suggested HIV testing but did not insist on this precaution. In addition, after the shot, the vaccine site could shed virus cells and cause illness (called "*contact vaccinia*") that could be dangerous to others.

Throughout 2003, just a few reported adverse reactions to the smallpox vaccine caused a drop in public participation. Unexpected heart symptoms (cardiac adverse events) were particularly alarming. Among the 250,000 soldiers vaccinated for the first time by March 31, fourteen (ranging in ages from twenty-one to thirty-three) suffered heart problems, either myocarditis or pericarditis or both. The 100,000 other soldiers being revaccinated reported no such problems with inflammation. Overall, the military was positive about the smallpox vaccine program.

By late March 2003, just under thirty thousand civilians had been vaccinated nationally, a small fraction of those expected to cooperate. Three of these volunteers had heart attacks, two of them fatal, and seven others suffered other heart-related problems. The CDC characterized the three first responders (ages fifty-five to sixty-four) who suffered heart attacks as already having "*clearly defined risk factors*" such as high cholesterol levels, cigarette smoking, and a previous history of heart trouble. The same appeared true for a fifty-five-year-old member of the National Guard who died of a heart attack five days after being vaccinated. The distinction between the vaccine's causing heart failure and its contributing to heart failure was lost on many. The threat of terrorists attacking with smallpox aerosol seemed more remote than these reported illnesses and deaths, especially as the Iraq war was declared over in May and the threat of bioterrorism faded from the news. Fifteen states immediately halted their vaccination programs. Many hospitals independently withdrew participation.

The federal government stayed committed to the program, although the intelligence data supporting this commitment, always vague, remained unspecific. In July 2003, the CDC was given $100 million to dispense to states to improve participation rates. The Johns Hopkins physicians who had organized the Dark Winter scenario remained optimistic about the smallpox campaign as a way to reduce the risks of bioterrorism.

Science for Biodefense

The Bioterrorism Act of 2002 was the comprehensive Congressional response to defend "*public health security*" and the strongest vote yet for technological solutions to the perceived threat. This legislation mandated increased surveillance of drinking water supplies and food and required secret stockpiles of drugs, vaccines (especially smallpox vaccine), and other emergency medical supplies. It defined the importance of "accelerated countermeasure research and development," that is, research on pathogens of potential use in a bioterrorist attack, with the Secretary of Health and Human Services tasked to fund research on them, in collaboration with the Department of Defense and the Joint Genome Institute of the Department of Energy. The Department of Veterans Affairs, with its large hospital system, was suggested as a biomedical research and development resource and was allocated million for this purpose. The bill affirmed that "accelerating the crucial work done at university centers and laboratories will contribute significantly to the United States capacity to defend against any biological threat or attack."

The distribution of funds in the 2004 Department of Homeland Security Budget ($29.7 billion total) signified the importance attached to technological innovation as a way to counter the threat of asymmetric WMD attacks on civilians. Despite the unsuccessful smallpox vaccination program, the largest single budget item was for Bioshield, the Bush administration's program for the development of better vaccines and drugs by pharmaceutical companies. Bioshield, a biomedical equivalent of Reagan's Star Wars program, promised universal protection from biological weapons but faced the uncertainty of the threat, the technology, and organization for national vaccinations or other campaigns. Pharmaceutical companies would receive $5.6 billion over ten years, with $890 million allocated for 2004. The budget allocated $455 million for technologies to counter biological, radiological, chemical, nuclear, and high explosives attacks on civilians. The budget included $40 million for biosensors in urban areas; $50 million was earmarked for the Metropolitan Medical Response System to enhance disease reporting.

Specifically to combat bioterrorism, the Bush administration directed important new funding to biomedical science. At the National Institutes of Health, the National Institute of Allergy and Infectious Diseases (NIAID), the largest research organization of its kind in the world and perhaps the most diverse and advanced in its projects, received $1.7 billion (of its total $4 billion budget) in new money to jumpstart the invention of new vaccine, antibiotics, and technologies for early diagnosis. Through Homeland Security, $70 million was also earmarked for eight to ten Regional Centers of Excellence located in major research universities and institutes. The number of Level containment laboratories for research on dangerous pathogens was also slated to increase.

Although NIAID would focus on protection against the best known select agents (smallpox, anthrax, tularemia, plague, botulinum toxin, and hemorrhagic fever viruses), Anthony Fauci, NIAID's director since 1984, claimed that important discoveries about infectious diseases in general could be made while investigating pathogens that posed little or no current public health threat. For example, the study of smallpox or Ebola might yield an antiviral drug that worked against other viruses, or research on the anthrax toxins might yield an antitoxin with alternate uses. Fauci compared the transition to counter-bioterrorism research to the development of AIDS research in the 1980s, when the public and young scientists had to be educated about its importance. He concluded that "there needs to be an accelerated and sustained effort to draw more competent scientists/researchers into the field of research on counter-bioterrorism."

The integration of basic science into the biodefense project meant that some university and medical center scientists would confront national security restraints not experienced since World War II or during the US offensive program before it ended with the 1969 Nixon decision. The emphasis would be on regulating pathogens, facilities, and scientists in the public realm. Mild restrictions were already in place in the 1990s; for example, the 1996 Anti-Terrorism and Effective Death Penalty Act required official reporting to CDC if listed pathogens were transferred from one laboratory to another. At that time, the presumption was that all biomedical scientists worked against disease. The Bioterrorism Act of 2002 required authorities at a research or medical facility to report work with any of the forty-two CDC select agents. The government would keep this bank of pathogen information secret, except in public health emergencies (presumably accidents, bioterrorist attacks, or other unusual outbreaks) or for congressional committees.

Authorities also had to submit the names of employees with a legitimate need for access to the pathogens to the secretary of HHS and the office of the US Attorney General. The names would then be run through criminal, immigration, national security, and other electronic databases to check for possible "*restricted person*" status. "Restricted persons" had been defined in the US Patriot Act of October 2001 (PL 107-56) as felons, fugitives from justice, those dishonorably discharged from the armed services, the mentally defective, illegal aliens, and foreign nationals from terrorism-sponsoring nations. By the new law, the attorney general could identify any individual as suspect who had been named in a warrant for violent criminal or terrorist activity or was suspected of spying for the military or intelligence operations of a foreign nation. These regulations applied to open nongovernmental laboratories, with the presumption that the government was monitoring its own classified research on dangerous pathogens.

To defend the nation against biological weapons—wherever they were and whoever had them—responsible biologists were asked to give up some measure of openness and free communication or choose other research paths. No matter what research they chose, some new technique or discovery might if published lend itself to an enemy's advantage. Quite suddenly, the entire enterprise of biology became potentially suspect. As early as November 2001 an editorial in the science journal *Nature* asked whether, with bioterrorism looming, the biological sciences had reached the end of innocence.

Between September 2001 and the end of 2003, the United States government developed an aggressive, nationalistic posture toward the potential threat of *biological weapons*. It invaded Iraq with the goal, among others, of eliminating its biological weapons threat; it created a new federal department for civil defense, with a considerable emphasis on biodefense; and it sought to conform an important segment of American biology to biodefense via pharmaceuticals and other technologies.

Alart on Potentially Dangerous Information

Paola De Castro and Federica Napolitani Cheyne have discussed about the bioterrorism. According to them: *biotterorism* has become a key subject in global communication, forcing the scientific community to urgently deal with problems of security and publishing of potentially dangerous information. This short note suggests that while editors of big science journals, supported by a regorous peer-review system, seem to be generally aware of security implications and how to handle them, greater attention to security issues should be paid in the editing of small science journals and of institutional report as well, where sensitive information is more likely to be published. Some practical suggestions are also recommended to help evaluate when the potential harm of publication outweighs its benefits and therefore a submitted paper should be modified or even not published.

Not a day goes by without the subject of *terrorism* being mentioned in the press, mass-media or other form of public health public communication. Themes and terminology such as bioterrorism, security concerns, potential misuse of scientific discoveries, weaponizing anthrax or small-pox have become common-place in our daily routine especially whenever an act of terrorism occurs.

The phenomenon of *terrorism* has become a key subject in *global communication*. At different rhythms it has forced Governments to implement security legislation, the general public to change or adjust their behaviour to new rules and procedures, intellectuals and historians to ask themselves about the causes of terrorism which is ever increasingly escaping from any form of standardized classification.

Some interesting questions arise: When did the scientific community perceive this phenomenon and understand that it can no longer consider itself exempt? In which way has or is it facing up to this? How important has the threat been considered at international level and more importantly at local level? What measures have been studied and implemented to contrast this threat? These are the questions upon which the scientific world is reflecting; delicate themes that academic and scientific societies along with those entities charged with public security are discussing in depth.

In addition, the online availability of most science journals as well as many institutional reports makes the protection of sensitive data even more difficult and the fact that online access to journal articles is often restricted (a registration or payment may be required) does not represent a major hindrance for potential terrorist looking for "*useful information*".

If Internet favours the dissemination of sensitive data and information, the editorial responsibility in publishing them is ever increasing.

Furthermore, institutional repositories are rapidly developing all over the world under the impulse of the *Open Access Movement*. The information contained in open access archives may represent an even greater threat in terms of security for various reasons: they can host preprints, self-archived material, articles which have not been evaluated or refereed, and last but not least their access is free and unrestricted.

Scientific Publication and Security

In February 2003 some researchers, representing prestigious universities and professional associations along with the editors of over 20 international science journals signed, under the name of "Journal Editors and Authors Group", a *Statement on Scientific Publication and Security* which was published simultaneously in many of the journals involved (including *Journal of the American Medical Association, Nature, New England Journal of Medicine, Science*) often accompanied by editorials and comments.

The *Statement* which is divided into four parts is preceded by a Preamble where the importance of the scientific publishing process is underlined: it allows and ensures the dissemination of that scientific knowledge which is crucial to the society since it improves the human condition in a myriad of ways. However "*new science*, as we all know, may sometimes have costs as well as benefits". Starting from here and under the urgency of dealing with terrorism the four parts of the *Statement* were conceived. These can be briefly summarized as follows:

1. Protect the integrity of the scientific process;
2. Understand the need to urgently face up to terrorism and security;
3. Find new processes for the control and review of scientific papers before their publication;
4. Avoid publication of papers if deemed by the editor to be potentially dangerous.

As was expected, this *Statement* raised considerable clamour and also negative comments. Principles of Ethics and academic freedom were called upon, the potential danger of "*censoring science*" was raised and it was also noted how the *Statement* basically failed to provide clear guidelines for the editors.

Regardless of any just consideration on the correctness and potential effectiveness of this *Statement*, undoubtedly it was a necessary step by the scientific editing community in answering terrorism and a first step in trying to lay the foundations of the undefined boundaries between freedom and security, between replication of experiment's results and potential misuse, legitimacy and harm.

Security Concerns

It might be necessary to again look at the aspects of *security in scientific editing* with greater incision. It may be even more important to pose the question about the publishing or not of papers containing "*sensitive*" information directly to the players not so much of the big journals (internationally known journals with high *Impact Factor*, IF) but the so called small journals, namely science journals distributed at national and even local level, still striving for their visibility in the scientific world.

It is these latter that should be educated, made aware and guided along these themes. This for at least two reasons:

1. The peer review system used by less prestigious journals is far less selective and rigorous compared to that in use by journals with higher refusal levels. E Wager *et al.* have defined this approach as "*bottom-up*". Their philosophy is to accept anything that "meets their minimum standard". Therefore it is more likely that articles with potentially dangerous information might evade

the control of the reviewer and/or of the editor, who often learnt their trade "*on the job*" and also work under the pressure of the "*publish* or *perish*" principle;

2. Original articles that report results relating to innovative techniques, important scientific progress or discoveries would unlikely be published in journals with no or limited IF, however it is in these that articles giving precise and detailed descriptions of events or places and known methodologies and techniques as to allow their reproduction even by those not trained could be accepted for publication.

Learning from a Practical Experience

The experience of the authors of this brief note have led them to ponder this issue. The recent publication of two articles within a short period, one in a quarterly science journal and the other in a series of institutional technical reports attracted their attention. Both publications are edited by the Istituto Superiore di Sanita (National Institute of Health, ISS, Rome - Italy).

The first article, on *risk assessment* in nuclear facilities was based upon a hypothetical severe accident occurring in a non-operational nuclear power plant or in other nuclear facilities (provisional radioactive deposits, research centres, spent fuel storage pools). The radiological impact and emergencies were also evaluated. Here one of the referees highlighted the paper's potential danger and suggested to revise the article, change the title slightly, avoid naming and localizing sites on the national territory and delete all those elements (including a certain terminology) which could attract the attention of malicious persons.

The second was a technical report, dealing with operating strategies and effective measures to prevent potential terrorist attacks through waterworks system. The report contained detailed information also on substances which could be hazardous to public health. Here it was not the referee (since the publication is not peer-reviewed) but the editorial staff who reported the question to the authors. A useful exchange of opinions followed between the researchers and editorial staff which led to a change in the title of the report aimed also at reducing the risk of retrieval by Internet surfers using search engines.

Some Useful Hints and Suggestions

Greater attention to security issues should be paid not only to small science journals but also to the so-called *grey literature*, which though not commercially published and generally printed in limited number of copies, is now often freely available online mostly through institutional websites. It contains specific and detailed information usually not subject to the peer review process and whose responsibility falls under authors, editors and issuing organizations. Even in the absence of peer review, however, the institution's responsibility cannot be disregarded. The editorial staff inside the institution or those who are responsible for the editorial policy should advise authors on the potential risks of spreading sensitive information. While big journals editors and reviewers are generally aware of security implications which lie behind the publication of potentially dangerous data, authors of grey literature or small journals often disregard them.

6

HOMELAND SECURITY

In the past, cities were walled and fortified to withstand attack from the outside. Settlement sites were selected not just for industrial purposes or for access to transportation as they are now, but because they were defensible. In today's world of bombers, missiles, tanks, and other mechanized combat equipment, the idea of fortifying a city seems to be obsolete. But this is not the case. Though walls may no longer do the job, modern defenses and security measures exist to harden critical infrastructure and lessen vulnerability to terrorist attacks.

Buildings are often engineered to withstand earthquakes and other predictable disasters (the World Trade Center towers were actually designed to withstand the impact of a small aircraft), but the issues of bomb resistance, air filtration, and threat detection are only modest concerns in the planning of most cities and structures. Excepting the construction of fall-out shelters, hardening against the threat of nuclear strikes has been considered totally outlandish and not worth consideration. This perception must change. In a world tainted by terrorism, the buildings likely to be targeted can and should participate in their own defense.

Although major office buildings have been the targets of the most deadly terrorist attacks in America, more critical infra-structure often goes virtually unprotected. The minimal monitoring of water stations, mail sorting offices, power distribution points, and even vital telecommunications facilities make them highly vulnerable to chemical or biological attack. Despite the government's emphasis on airport security, airports are not much better, and bridges and train stations have essentially no security of any kind. Finally, the people required to respond first if any of these places are threatened—firefighters, police, and paramedics—need better protection on the job. Lives depend on their having it.

As always, the best way to defend against attack is to prevent the attack from occurring. The FBI and other intelligence agencies have been doing an impressive job since 9/11 of predicting attacks and finding culprits, but they need tools that allow them to extract information without unreasonable intrusion innocent people's lives. These tools include better technology for detecting qualified threats and more effective decryption for intercepting communications. These technologies will require oversight, but that is no reason to delay their development.

Terrorist attacks are the kind of disaster that America now dreads most keenly. Unfortunately, it seems likely that another major attack will occur on American soil. Nanotechnology will help mitigate the damage caused by such an attack, though only changes in policy that are beyond the scope of this book can eliminate the threat entirely. Other types of crises unrelated to terrorism could also be addressed by nanotechnology. Fires, floods, natural disasters, and the outbreak of disease are all areas where nano-applications can make a significant positive difference. So nanotechnology for homeland security may have even more dual-use applications than nanotechnology for military applications and it is easy to see how it could directly impact all of our lives.

Hardening the Hearts of Cities

In recent years, the most harmful terrorist attacks on America have been explosive attacks on large buildings. The first World Trade Center bombing, the Oklahoma City Federal Building bombing, and the 9/11 attacks all involved the destruction of office complexes. Commercial landlords and civic authorities reacted to the 9/11 attacks by implementing cursory inspections for vehicles pulling into parking lots under high rise buildings and by imposing more vigorous campaigns to check visitors' photo identification. While these precautions may help, how can parking attendants hope to find bombs that may be concealed in a suitcase or even under the seats of an SUV? How many would recognize a bomb even if they found it? And what prevents a terrorist from forging a photo ID?

Some of the sensor technology could be applied to inspecting vehicles, but a bomb can also be detonated from the street in front of a building and a car containing a bomb may run into or through a barricade. Individually harmless chemicals can be brought into buildings on separate trips, then mixed therein to produce disastrous weapons. How can office and apartment buildings be hardened to reduce the impact of such attacks? Nanotechnology offers several solutions that could be integrated into the design of future, attack-resistant buildings.

Start by considering structural issues. In San Francisco and other key earthquake zones, buildings are designed to be strong but also flexible. Some designs utilize a flexible central mast to support the rest of the building. Constructing these masts of nano composite materials or using nanocomposites to replace today's ubiquitous structural steel and reinforced concrete could significantly improve buildings' performance and protection.

The structure of a nanocomposite material resembles that of a normal composite material such as reinforced concrete, but shrunk to the nanoscale. Reinforced concrete consists of ordinary concrete poured over a steel mesh called *rebar*. It is formed at the macroscale, but it still manages to combine the hardness and compressive strength of concrete with. The yield strength of the rebar to create a material with properties superior to either of its components alone. Nanocomposites could work the same way, but at the molecular level. For example, a nanocomposite might be made of a high-strength plastic wrapped around a rebar of nanotubes. Such a material could vastly outperform conventional construction materials, and a building based on it could withstand a much stronger explosion than anything that exists today.

Explosions typically cause damage in two ways. The first is through the blast concussion. The second is through heat, which is what actually caused the collapse of the World Trade Center towers. (The impact of airplanes alone would not have toppled the towers, but the immense heat generated by the burning jet fuel weakened key girders and supports.) Advanced nano materials can be used to deal with this problem. Buildings incorporating this and other nano materials now under development will have a much greater chance of surviving the heat of explosions and other kinds of fires.

Because of their immense strength, the amount of these nano materials required to build a nano-enhanced building is actually much smaller than the amount of conventional materials that would otherwise be used. This opens up whole new design possibilities and the nano-school of architecture may far be far away.

Smelling Smoke

Buildings and homes are already equipped with sensors for many common hazards. Almost all cities have codes requiring *smoke alarms* and *carbon monoxide detectors*. Even *burglar alarms* are a kind, of sensor, though for a very different sort of problem. Continuous environmental monitoring for chemical and biological weapons may never make sense for individual homes (though concerned citizens will certainly have the option to install such systems), but it makes a great deal of sense to put sensors in the air and water handling systems of large office and apartment buildings.

Most modern high-rise buildings already contain complex air handling systems for central air conditioning, heating, and humidity control. In many of these buildings windows cannot be opened for climate-control efficiency and safety reasons, but that makes them more vulnerable to attack. If a toxin were introduced into the air supply, getting clean air would require breaking the windows (causing another hazard in the form of falling glass), and clearing the building afterward would be extremely difficult.

If *nanosensors* were installed in high-rise air handlers and if the same kinds of filters we looked at for next-generation gas masks were introduced into the systems, the possibility for contamination would be reduced significantly A building could be evacuated if a toxin were detected and air could be safely filtered in most cases. During day-to-day operations, these detectors and filters would be low cost and low maintenance. They might also have the beneficial side effect of extracting ozone, particulates, and other common pollutants from the building's air. In this case, preparing for disaster would also mean improving air quality all of the time.

These same kinds of hardening processes could also be implemented for key infrastructure like water, electricity, sub ways, mail centers, and communications. In these cases, the upgrades arc not really optional. To fail to equip a water pumping and filtration station with state of the art sensor technology, for example, would be simply negligent. Remediation in the event of disaster would also have to begin at these points, so it makes sense to keep a supply of common remediation compounds on hand.

First Response

The term *first responders* has grown to mean police, firefighters, paramedics, and all other emergency workers who represent the front line of defense in case of a terrorist attack within the United States. These men and women face above-average risks in their day-to-day work, and the additional threat posed by terrorists both exacerbates certain basic problems and introduces new ones.

Like soldiers, police officers are at risk from enemy fire and often have to wear cumbersome protection including bulletproof vests while performing their duties. Firefighters require full sets of fire retardant suits, portable oxygen supplies, and axes, but despite these precautions even the best-protected firefighter can only remain in a burning building for a few minutes. Paramedics routinely deal with sick and contagious people, which is essentially a lower level biological attack. What can nanotechnology do to help?

Clearly the advanced-armor nanotechnology for soldiers and heat-resistant nanotechnology for buildings will help first responders. Chemical-defense nanotechnologies will be useful not only for their stated purpose, but also for use in filtering smoke and various toxic chemicals from building fires. Nanotechnology sensors may also be useful in replacing canine units for many purposes—even a dog's impressive sense of smell cannot detect a single molecule of something. Sensors will certainly have applications for diagnosing people and determining the threat level of a situation.

Nanotechnology also has some specialty applications for first responders. When such professionals discover a bomb, the protocol is to remove and safely detonate it if possible. This may require transporting a device and taking some risk of detonating it by mistake. To help reduce the impact of such an accidental explosion, nano materials like US Global Nanospace's Blast-X can be integrated into wall panels to significantly reduce the effect of an explosion. Such materials might also be used as blast retardants in trucks, carrying cases, and even specially modified body armor.

First responders also need remediation agents (chemicals that can break down nasty molecules or biotoxins) at least as much as the military does. One interesting solution to the problem involves using photo catalytic self-cleaning nanolayers (sometimes called *PSC layers*). These are surface coatings consisting of a nano-thin layer of material (sometimes something as simple as silver titania) that can break down various harmful contaminants when exposed to light. The National Technology Transfer Center's Emergency Response 'technology Pro gram claims that anthrax, smallpox, botulinum toxins, ebola, cholera, bubonic plague, nerve agents, mustard agents, hydrogen cyanide, tear gases, and even exhaust fumes and smoke can be broken down by exposure to PSC layers. These nanolayers could find applications in chemical-defense masks, air handling systems, and self-cleaning lighting, paint, glass, and other household fixtures. PSC technology has been developed by PPG, GE Lighting, Pilkington Glass and a few groups in Japan, where the idea has been championed by Akira Fujishima of the University of Tokyo.

There are other promising technologies for first responders. Many of them are very similar to the soldier nanotechnologies we have already discussed (although missing the active camouflage and weapons kit).

Clean It Up!

Decontamination following either an attack or an accident is a significant challenge in homeland defense. If prevention fails, first responders can deal with the immediate issues of safety and containment, but the threats of dispersed toxic chemical or biological agents require broad remediation and decontamination strategies.

We have already noted that chemical reactions occur at surfaces, and that *nanoparticles* have huge ratios of surface to volume because of their tiny size (half an ounce of nano-powdered alumina has more surface area than a football field). So *nanocrystals* and powders will react with toxic agents far more rapidly and effectively than traditional crystals or powders. This suggests that decontamination with such powders can be rapid and complete. The powders can either adsorb the toxic species (much like World War I gas masks that used activated carbon) or react with and chemically destroy the toxin (like the PSC layers discussed above). The powders can actually he sprayed on or applied with special mittens being developed by the Army Research Laboratory.

Decontamination of equipment, buildings, and even people is important enough that it already has a name (Decon) and some component serial numbers (M-11 and M-295) in the U.S. Army. Dual use, both for military and civilian protection, is a clear advantage of these nanotechnologies, because cleaning up the mess is important after an accident such as an oil or chemical spill, just as it is on a battlefield after a chemical or biological attack.

Human Repair

Since *nanoscience* works close to the interface between man and machine, some terms occasionally slip from one context to another.

Machines start to be described in human terms and humans to be described in machine terms. An excellent example of this, and one of the most promising fields of nano science, is human repair.

One form of human repair involves fixing "*mechanical*" problems with the human body, such as bone fractures, torn muscles and ligaments, burns, and cuts by helping the body heal itself. This differs from the conventional method for replacing bones using steel or ceramic implants, stapling tissue, and using other invasive tactics.

Human repair can greatly accelerate the process of healing. Consider the common case of a broken bone. In order to heal, it will probably have to be immobilized. This can be done with a plaster cast for many simple arm and leg fractures, but if the fracture is complex it may involve surgery bone nails, or even full-body immobilization.

It usually takes months for a bone to set and even longer for it to heal completely, but human repair can change this. Research done in Sam Stupp's group at Northwestern University addresses this very problem. Stupp's approach consists of injecting the site of a fracture with small molecules that assemble themselves into bonelike fibers that bridge a fracture. The self-assembly of these fibers from the liquid can rake just seconds, and as soon as it is in place, healing can begin. Osteoblasts, cells that aid in the development or repair of bones, adhere to the *nanofibers*, and natural bone forms to bridge the fracture. This approach to human repair can result in dramatically faster recovery

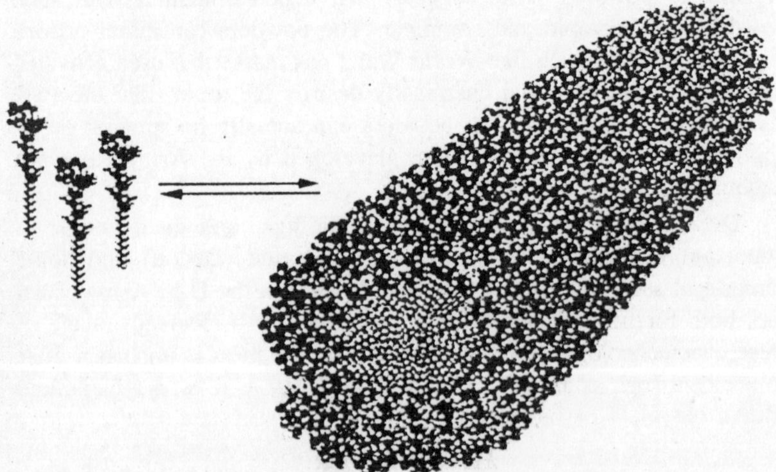

Fig. 6.1. Peptide amphiphile molecule (left) that self-assemble to form a cylindrical mandrel (right) on which bone can grow.

from injuries, less surgery, and a better mend when healing is complete. It also does not require steel bolts or implants that must be surgically removed later or that remain with a patient for the rest of his or her life. Almost everyone breaks a bone at some time. Wouldn't it be nice to be able to walk our of a hospital just days after being wheeled in with a broken leg?

Another form of human repair centers on controlling bleeding from traumatic injuries. This innovation will increase our ability to treat mass casualties, and it has obvious dual-use applications.

Information War

More and more of modern warfare is about who controls information, not just who has the most troops or the best weapons. The information war takes place at many levels, including intelligence efforts before a confrontation, identifying enemy positions, cyber-war fire, and protecting lines of communications. Almost all warfare now has an information component, but we will focus on two or three aspects in which nanotechnology is likely to change the game significantly.

High-performance Computing

The first and most obvious influence of nanotechnology will be in high—performance computing. It is clear that there is a limit to how much microchips based on current designs can be improved. Already the smallest components are entering the nanoscale, and as they do, engineers are confronting the strange properties of the quantum world. Most projections agree that by the year 2010, Moore's Law, the

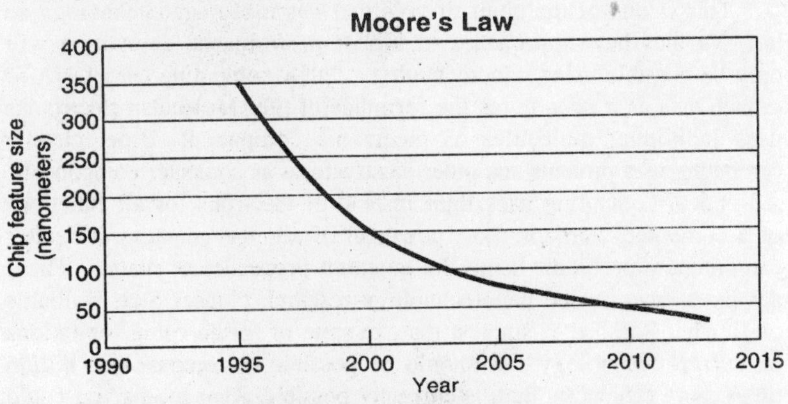

Fig. 6.2. Moore's first law predicts the rapid exponential growth of the density of transistors on a computer chip.

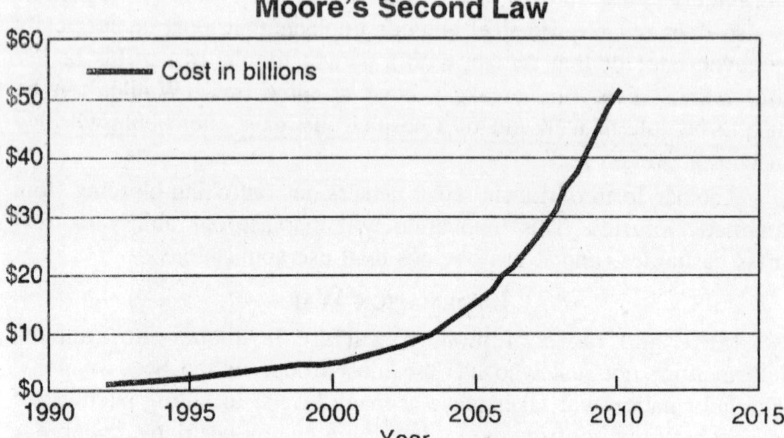

Fig. 6.3. Moore's second law notes the rapidly increasing price of fabrication facilities for making computer chips.

venerable computer industry maxim that computer components halve in size and double in performance every 18 months, will be end compound this, Moore's less cited second law that chip fabrication facilities double in cost with each chip generation shows no chance of abating. This means that by 2010 we might not only start seeing dramatically declining improvements in silicon technology but that the fabrication lines to manufacture the most cutting-edge chips will cost between $25 and $50 billion. At that pricey most semiconductor companies would go out of business.

This is one of the chief drivers that has made nanotechnology so hot. Various new approaches to higher performance computing will only be possible using nanotechnology. DNA computing uses DNA to crunch data as it now stores the formulae of life. Molecular electronics uses individual molecules as electronic component. Bioelectronic computing uses proteins and other biostructures as computer components. All-optical computing uses light instead of electrons for all functions of a computer. Perhaps most powerful of all, the quantum computer can compute problems using the quantum properties of matter. These advances have caused nanotechnology research pioneer Stan Williams of Hewlett Packard to suggest that, in spite of forthcoming limitations on current technology, "It should be possible to compute one billion times more efficiently than is currently possible. That means you could hold the power of all earth's present computers in the palm of your hand."

Faster computers are not just essential to the economy. They also allow for a number of security applications. For example, you often show your photo ID to enter a building. Devices exist that can instead scan your hand, eye, or face to determine your identity, check you against a list of restricted people and provide building security by letting you in or locking you out. These devices are much more secure than a quick inspection of a photo ID, but they have a problem: comparing data from a user to a small library is fairly easy, but if the library of users includes tens of thousands or even millions of people even the fastest computers can be swamped. The problem is even more severe if a DNA database of known terrorists is to be developed and used to help track them down. Although this data base effort is controversial, DNA for forensic use has been common for some years.

Image recognition in *satellite photography* is another potential nano application, especially since nanotechnology based computers are showing a particular penchant for pattern recognition. Data storage is yet another—DNA-based storage is one of the most efficient methods for long-term storage where access speed is not an issue, and three-dimensional "*holographic*" (light-based) storage offers quick retrieval as well as storage capacities much greater than anything we have today. Both of these data storage techniques exist already, though they are not yet commercially available. Nanotechnology has already made gigabyte hard drives reality, and that is proving to be a tiny fraction of what the tiny can do.

The loftiest goal of high-performance computing remains *artificial intelligence* (AI). AI would allow robots to take on many of the most risky jobs, such as serving in front-line ground forces. True, self-aware AI is probably still quite far in the future, but machines that can think their way out of complex situations may be just around the corner.

Cryptography

Code-breaking is essential to any serious intelligence gathering. If a nation wants to intercept terrorist communications or enemy transmissions in the battlefield, it must have good code-breaking ability Arguably the tide of World War II turned for the Allies when Alan Turing and his colleagues broke the Nazi cipher called *Enigma*. By 1942 the allies were decoding tens of thousands of Nazi communications each month, and the intelligence they accumulated has been estimated to have shortened the war by at least two years (though it is hard to know what effect the atomic bomb would have had). However, Turing's efforts took several years, an amount of time which is impractical in modern warfare and intelligence gathering. Today, advanced *cryptography* is widely available. Keys and encryption schemes can be changed quickly. Terrorists and hostile military forces take advantage of these technologies to make it difficult to recover information from captured computers or intercepted transmissions.

Many modern digital encryption schemes (such as RSA, a common scheme for e-commerce applications) rely for their security on the idea that large numbers are very difficult to factor and that the time it takes to factor them goes up exponentially with the size of the number. While a computer can in principle break these codes using a guess and check technique, breaking today's best commercially available codes using this approach would take a modern PC (circa 2004) more time than has passed since the fall of Rome. For this reason, the common commercial encryption schemes that permit data security within our society can also allow terrorists and other criminals to exchange data with almost perfect security.

Quantum computing is one solution that could turn the mathematics of code-breaking on its head. A quantum computer can, effectively, work on all the possible solutions to the problem at the same time rather than in sequence. In reality the solution is a bit more complicated than that, but there are reasons to believe that a single quantum computer could solve an encryption problem far beyond the reach of the mightiest PC—and it could do so in just a fraction of a second.

Breaking codes is more of a double-edged sword than most military or defense technology. The overwhelming majority of people using cryptography are employing it for lawful and decent reasons including trading stocks, buying goods, or protecting the privacy of their e-mail over public networks. Even more than most nanotechnology, decryption technology with this much power must be used sparingly and its abuse must be punished diligently. Quantum computing has applications far beyond decryption. Searching massive amounts of data and even modeling other quantum mechanical systems are two obvious examples.

Pervasive Computing

Relatively little thought has gone into the form, shape, or mechanical properties of modern computers. One or more silicon chips form the core of almost all electronics from computers to music players, and the silicon chips are still centimeter-scale pieces of silicon in plastic packages often trimmed with metal heat sinks. The chips sit on circuit boards of rigid plastic. This is a perfectly good way to build electronics for many applications, but it prevents them from being integrated into clothing, packaging materials, wall paint, and many other places where they could be useful.

It may not be obvious why it is desirable to put computers into wall paint. Many people justifiably feel that computers are already too much a part of their lives, but the applications for tiny, flexible computers (so-called "*pervasive computing*") are very intriguing. For example, pervasive computing could have household applications, such as controlling lighting, temperature, or music volume based on inputs ranging from what you are doing to the expression on your face.

In addition, pervasive computers have security applications when coupled with appropriate sensors, since they cannot be disabled easily the way motion detectors can. Unlike most security systems, which operate best after business hours, they can so allow the continuous monitoring of an area even during periods of normal activity. Finally they are necessary for many of the soldier applications we've discussed, including the next uniform.

So far the most promising avenue for this kind of capability is a nanotechnology called *molecular electronics*. Molecular electronics involves using individual molecules as components in electronic circuits. This is the ultimate level of miniaturization for electronics of the type that we are used to transistors, capacitors and the like. In addition to being very small, molecules (as opposed to metals or crystalline materials like silicon) are generally fairly soft and flexible. Molecular components may not need to be affixed to a rigid plastic board the way current electronics do. Instead, it may be possible to put them on softer polymer surfaces or other flexible backplanes. This would allow them to he used for many kinds of computing applications.

Molecular electronics will initially be soft and cheap. For example molecular thin film transistors (and we mean *thin*—just a few nanometers in thickness) now being developed by Phillips, Lucent, and others, are designed primarily for disposable applications like identification tags to label soup cans, pieces of mail, or even individual pills. The soup and pill labels would help track possible contaminations, including terrorist modifications, and allow fast recall. Recall the concerns with meat product contamination and cyanide in Tylenol bottles.

System Diversity and Survivability

One of the greatest threats to American security is the vulnerability of our computing and communications infrastructure. This comes in several forms. The fact that a single computer architecture (the PC) and a single operating system family (Windows) now comprise the overwhelming majority of our data systems represents a huge threat, since any weak ness discovered in either design may be exploited on millions of other systems. This problem is similar to dependence on a single crop for food. For example, in the Irish potato famine a blight wiped out the whole crop of potatoes and there was no alternate crop, such as corn or wheat, to see the population through the crisis. High-impact computer viruses have shown that the analogy holds for computer systems. What often starts as a prank can result in billions of dollars in damage and lost productivity. While nanotechnology is not currently poised to do anything about operating systems directly, it may help to evolve the underlying architecture of systems, diversify them, and make them more secure.

As mentioned earlier, nanotechnology is already promising at least five possible major revisions of traditional computer architecture. These include all-optical computing, bioelectronic computing, DNA computing, quantum computing, and molecular electronics. Some of these, like all-optical computing and molecular electronics, are general purpose and may work quite similarly to current machines. Others, like DNA and quantum computing, will be much more effective for speeding certain kinds of computations (such as code breaking) and can offer some types of tamper-free computing, but may never find application in general-purpose computing. All of them change some basic rules.

"While we have talked a lot about chemical and biological weapons, we have not spent much time discussing the third major category of weapons of mass destruction—nuclear weapons. There are many reasons for this. Not even nanotechnology will protect a city from a megaton nuclear device. It may be helpful in detecting the weapon or even shooting down a missile carrying one, but there are no technologies on he near-term horizon which will do much to protect us if the worst happens.

Having said that, a nuclear weapon from a less-developed state is unlikely to be an advanced, hydrogen bomb with a megaton yield. Smaller, so-called. "*tactical*" or "*dirty*" nuclear weapons are much easier to make, buy, or smuggle into a tar get zone. They cause a great deal of immediate damage, but there are other environmental effects of a nuclear detonation, which must be dealt with similarly to remediation of chemical and biological weapons. Another effect of such weapons is the nuclear *electromagnetic pulse* (EMP). An EMP is a wave of massive energy that can disable or destroy electronics well beyond the reach of the actual nuclear blast. The military specification for electronics requires ruggedness well beyond most civilian specifications. Not only must military electronics operate in hotter, colder, and damper climates than civilian electronics, they must also have some resistance to EMPs. While it is possible to improve the resistance of electronics to EMPs, it is very difficult to shield them completely, and most shielding requires significant sacrifices in terms of weight and performance. There is reason to believe that all-optical computing and DNA computing as well as some other nanotechnology-based computing options will be naturally resistant to EMPs. They will also have better performance in other hostile conditions including outer space.

All-optical computing (using light rather than electricity to process data) is one example that will not only resist EMPs, but will also be less sensitive to electronic eavesdropping. Since the optical signal is confined to a fiber much more tightly than an electronic signal is truly confined to a wire, it is harder for an external piece of equipment to listen in. The limiting factor to creating an all-optical computer is the design of a small, efficient optical switch to perform the same duty as a transistor. Bulkier optical switches for all-optical communications net works exist, but optimization to the point of usefulness for a complete computing platform has not yet been done. It will be done eventually, and all of the most promising designs to date are based in nanotechnology.

All of the nanotechnologies that we have just surveyed for homeland security exist in some form. Some are already commercial products and some are still several years from development. Even the far-out notions such as quantum computing are close enough to hand that corporations, venture capitalists, and universities are becoming interested. For example, D-Wave, a Canadian startup interested in practical quantum computing, raised a large round of private equity under the most difficult conditions at the beginning of 2003.

There are dozens of other ways in which nanotechnology could affect homeland security, and many of them will only be recognized as fundamental research continues. Unlike defense technology, however, homeland security technologies will impact all of our lives directly. Some, such as human repair, will only change life for the good. Others, such as advanced cryptographic technology, could be much more threatening. And still others, like Al or pervasive computing, could change our day-to-day lives as much in the next 15 years as electronics, the Internet, and telecommunications have done in the last 50.

7

Threat of Biological Weapons

"Weapons of *mass destruction*" is a terrifying term. We all have mental images of the horrors of a nuclear attack, and photos of Kurdish and Iranian casualties of Iraqi chemical attacks attest to the devastation of *chemical weapons*. The third weapon of mass destruction—the *biological weapon*—has been around at least since the Middle Ages when soldiers catapulted the bodies of dead *smallpox* victims over fortress walls in the hope of infecting their enemies or at least demoralizing them. Lately, *biological weapons* have been appearing in the news with increasing frequency.

The anthrax threat in Las Vegas in February of this year is an example. Surplus stores in Las Vegas sold out of gas masks, and talk-radio shows were swamped with callers asking about evacuation points. That threat turned out to be a false alarm, but the next one might be real. *Biological agents* are of concern in part because of the ease with which many of them can be manufactured, transported, and dispensed. And because of the lag time between a biological attack and the appearance of symptoms in those exposed, biological weapons could be devastating.

Many biological agents are contagious, and during this lag time, infected persons could continue to spread the disease, further increasing its reach. Hundreds or even thousands of people could become sick or die if a biological attack were to occur in a major metropolitan area. With the knowledge that several nations have produced and perhaps also deployed biological warfare agents, Congress in 1996 passed the

Defense Against Weapons of Mass Destruction Act, which authorizes the Department of Energy to establish a Chemical and Biological Weapons Nonproliferation Program. Under this and similar programs, Lawrence Livermore and other laboratories and institutions are working together to increase this country's capabilities to detect and respond to an attack by biological or chemical weapons.

Beginning as recently as Fiscal Year 1996 with a Laboratory Directed Research and Development strategic initiative, Livermore has rapidly expanded its chemical and biological nonproliferation program and is now playing a lead role in this effort, particularly as it pertains to defense against biological weapons. The Laboratory is applying its investment in biological science, engineering, microtechnology, computer modeling, systems analysis, and atmospheric science to a number of programs designed to improve the country's response to a biological attack. Personnel from departments and directorates across the Laboratory are at work on:

1. Advanced detection systems to provide early warning, identify populations at risk and contaminated areas, and facilitate prompt treatment.
2. Biological *forensics technologies* to identify the agent, its geographical origin, and/or the initial source of infection.
3. Methods for predicting the transport of biological agents in urban environments and for assessing the area and duration of the hazards associated with a biological attack.
4. New *decontamination technologies* to clean and restore facilities without causing further environmental damage.

Livermore is working closely with the U.S. military, various government agencies, and such major cities as New York City and Los Angeles to ensure that the results of these biological nonproliferation efforts meet the needs of military troops, the FBI, local law enforcement personnel, fire fighters, public health officials, and others who would likely be first on the scene following a biological attack. Together these groups are answering questions to help create the best, most task-appropriate, and most usable system possible. For example, how accurate do sensors have to be? What level of false alarms can be tolerated? Where will sensors be located—in buildings, on emergency response personnel, or at other sites? How much training will be feasible for emergency response personnel on the use of sensors and decontamination agents—that is, how user-friendly must these processes be? Livermore is developing a strategy for defense against the use of biological weapons that integrates technology, operations, and policy and provides a framework for coordinated local, state, and federal emergency response.

Better Detection Systems

A key factor limiting the nation's ability to protect against a biological attack has been the state of biodetector technology. Only now is technology becoming available that permits identification of biological organisms within minutes, when concentrations are low but often still dangerous. Before the revolutions in *genomics*, *biotechnology*, *microengineering*, and *microcomputers*, such identification could only be done in a laboratory and took days to weeks. Soon, however, technology advances—many of them made at Lawrence Livermore—will offer the possibility of rapid, accurate, and *sensitive biodetectors* for use in battlefield or urban settings.

Automation is Key

Livermore is developing two types of fully *automated biodetectors* for realtime sample collection, detection, and identification in the field. A *miniature flow cytometer* (known as *miniFlo*) uses an immunoassay system to look at the proteins and other material on the surface of cells, and a portable PCR (*polymerase chain reaction*) unit identifies the DNA inside the cell. Because of their small size and efficiency, both units process data much faster than their laboratory-scale cousins, while maintaining the highest level of sensitivity. To fully automate sample collection and preparation, Livermore is developing and testing components for an aerosol *biocollector* and a *microfluidic* sample preparation system. The device will collect and sample particles in the air, including biological agents, if present.

To maximize detection potential and give faster results, the PCR unit and miniFlo are also being "*multiplexed*" to handle multiple samples at once. Other system improvements are being made to both instruments to lower the rate of false positives (*false alarms*), increase the sensors' sensitivity, and make the systems even smaller, more rugged, and less reliant on consumables than they are now. Livermore expects to have continuously operating, integrated biosensors available for use within the next few years. With two types of sensors working in *tandem*, the chance of false alarms will be reduced considerably.

Tolerance for false alarms differs greatly for military versus civilian situations. *Deployed troops* are already in a state of heightened readiness, with protective equipment available and the training required to react to attack situations. In contrast, with civilians, false alarms could lead to injuries and perhaps to dismissal of future legitimate alarms. Thus the military may be able to afford some level of false alarms, but the goal for the civilian sector is no false alarms.

The *miniFlo* and the *PCR systems* have proved their mettle against established performance criteria at the U.S. Army's international Joint Field Trials at the Dugway Proving Grounds in Utah. At Dugway, participants use a variety of instruments to detect simulant materials representative of typical biological weapon materials. At the 1996 Joint Field Trials III, miniFlo was superb at detecting *Bacillus globigii* and *Erwinia herbicola* (nontoxic simulants for anthrax and plague respectively) at various low concentrations. Overall, miniFlo detected 87% of all unknowns with a false alarm rate of under 0.5%. At the 1997 Port/Airbase Advanced Concept Technology Demonstration and the January 1998 Dugway Joint Field Trials IV, the portable PCR unit clearly demonstrated the potential of PCR as an effective technique for field identification of DNA.

Networked Detectors

A networked system of these or other *biodetectors* could provide U.S. troops in the field with early warning of a biological attack. That is the goal of a project for the Department of Defense known as JBREWS (Joint Biological Remote Early Warning System), on which Livermore is collaborating with Johns Hopkins Applied Physics Laboratory and Los Alamos National Laboratory. JBREWS will consist of a network of sensors and communication links. By tying this network into the military's existing communications systems, JBREWS will take advantage of well-established command and communications procedures. Initially equipped with commercially available sensors, JBREWS is being configured so that improved biodetectors can be incorporated into the system as they become available.

Livermore is responsible for what is known as "C4I"—*command*, *control*, *communications*, *computers*, and *intelligence*. The Laboratory is developing the connectivity between the sensors and the control station, the software for all sensors, and an automatic analysis and reporting system that runs up through the military chain of command. JBREWS is scheduled to be demonstrated in a Department of Defense Advanced Concept Technology Demonstration in 1998.

Biological Forensics at Work

If a bacterium or spore appears in a collected sample, how will a biodetector know what it is? The key to identification will be a library of "*signatures*" of the makeup, function, and DNA of various biological agents that will be stored on a *microchip* in the detector, together with pattern-matching software and code for reporting results. This technology will allow advanced detectors in the laboratory and ultimately in the field to quickly match the signatures of collected particles to signatures in its memory, in much the same way that fingerprints are matched.

Building on years of experience in *genomics* and *biotechnology*, Livermore scientists are expanding the information base of the DNA sequences of biological agents to enable rapid, unambiguous identification of biological agents. To facilitate this process, they are developing ways to speed up the process of finding unique DNA sequences among organisms. A process known as *representational difference analysis* helps to identify unique DNA sequences. Parts of the DNA of two organisms are mixed. If they stick together, they match; if they do not stick, they are unique parts.

Currently, this process is cumbersome and slow, but Livermore scientists are working to automate it to be able to examine many sequences in parallel. Another project is studying specific pieces of bacterial DNA and examining the possibility of using their location as an indicator of differences among strains. A third project is *investigating virulence factors*, which are the genes that give a biological organism its infectivity or toxicity. If a *bioweapon* is being genetically engineered, those genes might be moved to an unnatural host in an attempt to thwart detection and identification.

In addition to identifying the particular agent being used, tools being developed at Livermore also seek to provide information that will help to identify the perpetrator of a biological attack. Livermore biomedical researchers were among the first to study regional differences among the various naturally occurring strains of *anthrax* and other biological agents. Law enforcement personnel will be able to match data about a pathogen with data on regional or strain characteristics and with data on worldwide biological research, epidemiology, and infectious diseases and respond to the threat.

Predicting Agent Dispersion

The ability to accurately predict the dispersion, concentration, and ultimate fate of biological agents released into the environment is essential to prepare for and respond to a biological agent release. Of particular concern is the threat to civilian populations within major urban areas where potential terrorist incidents are more likely to occur.

There the hazard from a biological-agent release could be confined to a localized area within or around a single building or extend out to a large portion of the city or even into the surrounding suburbs, depending on the particular agent release, the quantity and duration of the release, and the meteorological conditions under which dispersion of the agent occurs.

Computer simulations of biological releases are critical to the design and placement of *biosensor systems*. They also aid in risk assessment, disaster planning, and emergency response training. If a biological release were to occur, real-time predictions of agent concentrations would be used to characterize the source, estimate exposure levels, identify affected areas and best evacuation routes, and later assist with decontamination. Accurate information about the likely course of a bioagent attack is key for emergency response managers, who must notify health officials, inform emergency response teams, and make public safety decisions.

The urban biological release problem is quite complex and requires modeling capabilities that are still in the early stages of development and application. For example, models of airflow inside buildings and subways have been developed to some degree but do not accurately incorporate the decrease in airborne concentration that results from deposition of the toxic material on walls, ceilings, ventilation ducts, and other interior surfaces.

Similarly, computational fluid dynamics models of the highly distorted flows and dispersion patterns created by complexes of buildings are just beginning to include the effects of *biological aerosols* (gravitational settling, deposition, and viability degradation) and multiple building interactions. Lawrence Livermore, Lawrence Berkeley, Los Alamos, and Argonne national laboratories are working together to develop an integrated and validated atmospheric modeling capability for biological agent releases in an urban environment. They will be applying these models to case studies in a range of release scenarios, from closed office buildings, to subway systems, to stadiums and street corners. The goal is to make the models applicable to real-life situations and ultimately to integrate them into the incident response capability of the National Atmospheric Release Advisory Center, located at and operated by Lawrence Livermore.

Decontaminating a Site

After an area has been exposed to a *biological attack*, it must be decontaminated before it can be reopened to the public. Livermore and Los Alamos national laboratories are working together to develop decontamination strategies for three scenarios—an open stadium, a semi-enclosed subway, and an enclosed area such as an office or home. Certain decontamination methods might be acceptable for one scenario but not another. For example, more corrosive reagents and large volumes of water might be acceptable in a stadium but could not be used in an office building.

Plain household bleach is one of the best *decontamination agents* around, and it is used regularly in biological laboratories throughout the country. But 5% sodium hypochlorite (as bleach is more technically known) is a very caustic product, so it must be used with care. The team is working to develop decontamination methods that are as effective as bleach but more acceptable environmentally.

Decontamination proceeds in several stages, from cleanup of gross contamination such as puddles of agent, to localized decontamination of walls or furniture that were directly exposed to the agent, to cleanup of ductwork or inaccessible cracks for hidden contamination, and finally to long-term remediation such as special paints or sorbents to destroy small quantities of agent that are left after completion of other decontamination. These stages may require different cleanup materials. A variety of liquids and powders are being studied, as is an array of delivery methods such as foams and gels. One treatment method that has been found to be effective and more environmentally acceptable than *hypochlorite* is *peroxymonosulfate*, which is an acidic oxidizer.

The treatment of a simulant for *anthrax* with these oxides can easily be compared. The selected method must be not only effective but also easy to use with minimal training. The social and political issues involved in decontamination and reentry to a site are not being overlooked. Central to these concerns is "How clean is clean enough?" The team is coordinating with the biosensor developers to devise sampling and analysis systems that can verify that decontamination is complete. One hurdle for the decontamination process is that no real-time biodetector currently under development at Livermore uses an assay that can distinguish between viable organisms and dead or decontaminated ones. Work has begun on a "*viability assay*" based on flow cytometry to provide this important piece of information so that decontamination can proceed in a timely manner.

Responding to the Threat

The threat of biological weapons is all too real, and the U.S. must be prepared to respond if a bioattack occurs on the battlefield or in a civilian setting. During the 1991 Gulf War, the U.S. had no systems available for rapid, timely field detection of bioagents. The situation today is very different. The military has deployed *Biological Integrated Detection Systems* (BIDS), which can tentatively identify the presence of a suspected biological agent in the field and warn soldiers to take appropriate action to protect themselves against the agent, pending positive laboratory identification. And there are also programs such as Livermore's that include new detection, identification, atmospheric modeling, and decontamination capabilities, which, combined with work by others on better vaccines and medical treatment, are bringing the country to level of preparedness that can meet a biological threat.

Livermore's New Biodetectors

Portable PCR

In late 1996, Lawrence Livermore delivered to the U.S. Army the first fully portable, battery-powered, real-time DNA analysis system. DNA analysis required many copies of a DNA sample, which are made by the polymerase chain reaction. PCR requires repeated cycles of an aqueous sample being heated close to the boiling point and then cooled. To detect DNA in a sample, a synthesized DNA probe or primer tagged with a fluorescent dye is introduced into the sample before it is inserted into the heater chamber. Each probe or primer is designed to attack to a specific organism—anthrax, plague, etc. If that organism is present in the sample, the probe attaches to its DNA. By measuring the sample's fluorescence, the instrument reports the presence (or absence) of the gargeted organism.

In Livermore's portable unit, the thermal cycling process takes place in a micromachined, silicon heater chamber that has integrated heaters, cooling surfaces, and windows through which detection takes place. The PCR reaction and DNA analysis take place in a disposable polypropylene reaction tube inserted into the heater chamber. Because of the low thermal mass and integrated nature of Livermore's silicon heater chambers, they require very low power and can be heated and cooled much faster than conventional units. So the unit is not only portable but also much faster and more energy-efficient than bench-top models. A multiple-chamber unit that allows the examination of many samples at the same time has been field tested.

MiniFlo

Livermore's miniature flow cytometer is the latest in a series of flow cytometers developed over the past two decades in Livermore's

Biology and Biotechnology Research Program Directorate. Flow cytometers are used in laboratories to analyze cells and their features, perform blood typing, test for diseases and viruses, and separate out particular cells or chromosomes. What sets miniFlo apart from other flow cytometers is its small size, portability, and sensitivity.

These features are made possible by a novel system that cases the alignment and increases the accuracy of flow cytometry. In a flow cytometer, the cells flow in single file in solution while the experimenter directs one or more beams of laser light at them and observes the scattered light, which is caused by variations in the cells or DNA. Instead of using a microscope lens or an externally positioned optical fiber as a detector, this method uses the flow stream itself as a waveguide for the laser light, capturing the light and transmitting it to an optical detector. This approach not only eliminates the alignment problems that plague traditional flow cytometers but also collects ten times more light than microscope lens does. Simpler alignment and more light mean better, faster analysis.

Bacteria are large enough for individual detection in the miniFlo, but viruses and proteins are not. So beads large enough to be detected are coated with an antibody and added to the sample. The virus or protein attaches itself to the bead and can then be detected. When different beads are coated with different antibodies, simultaneous detection of several biological agents is possible.

8

CHANGING FACE OF THE WAR

America possesses the strongest military in the world. In terms of manpower equipment training, and capabilities its armed forces have distinguished America as perhaps the only remaining superpower. This military might has been crucial to America's prosperity—the possession of great wealth requires the capability to defend it. It has also been pivotal to the geopolitical stability of the world. After the invention of nuclear weapons in 1945 and the Cold War that followed it, America, Russia, and the other nuclear powers developed large, robust arsenals that could not be eliminated with a single preemptive strike. Any attack on one of these nations would, in principle, lead to nuclear retaliation and to the total devastation of all the parties involved. This military doctrine, sometimes called Mutually Assured Destruction (or ironically, MAD), dominated strategic thinking for much of the Cold War. It also meant a shift in the methods and locations of war since another world war is unthinkable to sane leaders in the context of MAD. Instead, as JFK pointed out, the theaters of war changed to Asia, the Middle East, Africa, and other less-developed areas.

With the collapse of the Soviet Union and the end of the Cold War came a corresponding change in the threats to America. Most Americans (and citizens of other developed countries) are no longer concerned about invasion, as we were until World War II, or about extinction, as we were during the Cold War. In this era of globalization, we are concerned about terrorist strikes that result in economic or political instability as well as death. It's useful to look at why this shift occurred and at the change in thinking that has accompanied it in order to understand the new challenge to America's armed forces.

While this discussion is necessarily simplified and highly abbreviated, it still adds context to the issues.

During the Cold War, America and the Soviet Union were reluctant to fight or even skirmish with each other directly since matters might escalate out of control. Instead they tended to arm and train local militias, insurgents, and armies in the nations they considered important to their national interests, such as Vietnam, Afghanistan, Cuba, Korea, Iran, and Iraq. Indigenous troops were taught how to use modern weapons systems and how to fight guerrilla wars. The super powers also tended to back governments of convenience, generally defined as those that would be amenable to the superpower's own political or economic interests rather than those that were legitimately elected. This led to ongoing strife and a culture of violence that has caused many of these nations to remain hot spots to this day. While terrorism predates the Cold War, the overwhelming majority of today's most notorious terrorists, including Osamabin Laden, either come from hot spot countries or were trained there. These people are under no illusions that they are capable of fighting the United States directly to achieve their objectives. Instead, they are trained in unconventional war and, like good martial artists, apply force precisely at their enemy's weakest points often using his own strength against him. This means killing civilians, toppling symbols and monuments, disrupting the economy, and where possible using the other side's own technology.

During the world wars and all preceding conflicts, victory was paramount. Domestic casualties were minimized when practical but a fairly high casualty rate was considered acceptable. Enemy casualties, both military and domestic, were hardly given notice. This was the psychology that led to the firebombing of Dresden and to the nuclear destruction of Hiroshima and Nagasaki. In the Korean War, and to a much greater extent in Vietnam, this began to change. Americans questioned why troops were being sent for dubious purposes of a faraway nation that most people had never heard of. The human losses became unacceptable. Enemy casualties have also now become a major issue when fighting in countries that sponsor terrorism. This is attributable to a few causes. Most such countries are not representative governments and many feel that the citizens should not suffer for the sins of their leaders. Also, some feel the pangs of a guilty conscience, for the United States helped to create many of the situations that have led to the surge in world terrorism. While very few question that a country is justified in striking back against an attacker, the debate over the recent war in Iraq (especially Germany's position) has underscored the philosophical division on the policy of risking additional civilian deaths in an effort to stop terrorism.

While these changes are certainly profound perhaps the greatest changing influence on the face of war is not directly geographic political or psychological but economic. The spread of free markets as well as advanced technology such as transportation systems, computer networks, and telecommunications systems have resulted in an economy that grows more and more globalized. In 2003, Germany, France, and Russia all had financial interests in Iraq, which made their opposition to the war seem as much like economic pragmatism as true idealism. American policy was equally stigmatized by accusations that oil needs rather than security concerns were behind its actions. More generally, the economies of the developed world are linked so closely that a ripple in any one affects the others. Pain in South America has echoes in Russia and in Europe. A recession in America or Japan brings down the rest of the world. Free-trade regions that lower the barriers still more are cropping up in Europe, the Americas, Asia, and possibly even Africa.

America has the strongest economy as well as the strongest military on earth, so it must play its cards carefully. It is clear that war and instability are counterproductive to the global economy. Threats of conflict depress consumer spending, travel, international trade, investment, and stock prices, leading to bankruptcies, negative growth, and unemployment. These issues are of surpassing importance to America, which is a large part of why the first President Bush failed to win reelection even after prosecuting a successful campaign in the Gulf War. This has led military thinkers such as General John Sheehan to speculate that the core mission of the U.S. military is now not just to protect America, but to help maintain global stability for economic development. Many military leaders, including General Colin Powell, have become more dove-ish than their political counterparts, and it has already been largely forgotten that the cabinet post now called Secretary of Defense was not long ago called Secretary of War. Unlike in the days of empire, war no longer pays.

These elements form the challenge to American military, if at all possible conditions leading to war, such as the accumulation of weapons of mass destruction by rogue states or those that sponsor terrorism, must be detected and stopped before they become problems. When wars do happen, they will be fought in faraway places and harsh environments such as swamps, jungle and deserts. Neighbouring countries may not he friendly, so operations may need to be headquartered far horn the front lines. American and civilian casualties are unacceptable. Enemies are skilled guerrilla fighters who are relatively well armed, don always wear uniforms or move in units, and have made extensive use of unconventional weapons including chemical and biological ones. These enemies also have very few constraints: they are willing to use any weapon and any tactic necessary and they seldom worry about casualties on either side. Finally, the wars must be short to minimize economic consequences but peacekeepers may need to remain in the fields for years after a conflict.

This was, more or less, the scenario of the 2003 war in Iraq. Add to it the continuous press coverage that largely eliminates cheating or secrecy, and the Iraq mission seemed impossible even with DoD's half a trillion dollar budget. To its credit, the military stepped up to the plate and embraced the challenge. It made many changes to its organization such as relying more on intelligence and special forces groups, improving communications at every level, unifying command 0 domestic defense under the Department of Homeland Security, and increasing its capabilities for working in small, highly coordinated units. These changes are beyond the scope of this book, but they are an important first step.

The military was also very lucky. Despite concerns, no weapons of mass destruction were used—even found—during the offensive in Iraq. If they had been, the situation might have turned out quite differently. Yet this is far from the only example of how vulnerable even the strongest military in the world can be in the face of its new challenges. The solutions to many of these problems are technological, not organizational. The core research operations of the Department of Defense, including the Defense Advanced Research Projects Agency (DARPA) and the Office of Naval Research, have issued lists of Grand Challenges. These are the crucial technological hurdles that must be overcome for the military to fulfill its new mission. Nanotechnology is key to the realization of almost all of these challenges, and the rest of this chapter is dedicated to describing many of them and explaining how they are being addressed.

Agents of Mass Destruction

The use of chemical and biological agents as weapons of mass destruction is not a new phenomenon. People have been using biological weapons ever since they started throwing decomposing corpses and feces over city walls during sieges in hopes of causing the plague. Some of the first chemical weapons were poisons used on arrowheads. Chemical and biological weapons date back to prehistory.

Despite their early use, chemical and biological weapons never gained much popularity until modern times. Chemical and biological agents are generally difficult to "weaponize" or make into devices suitable for use in combat. Before missiles and bombs were common, this was a huge problem, since a plague represented a threat to attackers as well as to defenders when troops were closely mingled. In addition, lingering contamination would remain a problem for citizens and occupiers after the battle was over. Also, a change in wind or weather could blow a toxic cloud back at an attacker, and a storm or rain shower might reduce or negate its effects. Without a mechanism for delivering them to an opponent's homeland, chemical and biological weapons have virtually no value as weapons since they are so volatile and unpredictable. It is hard enough to protect an attacking army from such weapons. Until now, it has been virtually impossible to protect a civilian population but nanotechnology may change this.

World War I, which saw the first large-scale use of aircraft and modern field artillery, was also the battlefield on which chemical and biological weapons truly reached maturity. Blister agents (also called mustard gas), nerve gas, and a litany of other horrors killed troops and civilians in unprecedented numbers. Today, toxins such as ricin and sarin have been developed to be deadly in quantities of much less than a gram. To put this in perspective ricin is considered twice as toxic as cobra venom. These agents are ideal weapons for terrorists since none of the factors that make them poor military weapons apply to terrorist tactics. Terrorists do not consider the unpredictability of these weapons to be detrimental since their psychological effect is undiminished. They are not concerned about weapon zing them since they can easily use suicide bombers to attack people or to plant toxic compounds directly into a building's ventilation system or a city's water reservoir. Sarin was used in this way when the Aum Shinrikiyo Supreme Truth cult distributed it on the Tokyo subway in 1995, killing a dozen people and injuring thousands. Ricin also has a history—it was a ricin pellet in an umbrella tip that was used in the 1978 assassination of Georgi Markov, the Bulgarian writer and journalist.

Chemical weapons are easy to make from common materials. Ricin, for example, is derived from castor beans with the aid of some basic chemistry apparatus, all of which are avail able in any high

Table 8.1. The relative toxicity of some chemical weapons and biotoxins.

Agent	*Lethal dosage (LD 50 g/kg body weight)*	*Source*
Botulinum toxin	0.001	Bacterium
Diphtheria toxin	0.10	Bacterium
Ricin	3.0	Castor bean
VX	15.0	Chemical agent
GB	100.0	Chemical agent
Anthrax	0.004-0.02	Bacterium
Plutonium	1	Nuclear fuel

school science laboratory. The British authorities have already caught suspects trying to make the substance in London and there is some evidence that it was recently tested in northern Iraq. Hydrogen cyanide (HCN) is another easily obtained chemical agent. It is mass produced for industrial uses such as tempering steel, dyeing, engraving, gold mining, and the production of some plastics. It is also the active ingredient in Zyklon B, the gas used in Nazi gas chambers. It is still used in parts of the United States for criminal executions.

Unlike explosives, almost all of which are based on materials with a chemical similarity (they contain nitrates), there is no single commonality among all chemical or biological weapons. Even nuclear weapons are fairly easy to detect via the radiation they emit. Radiological contamination can be accurately and quickly measured using a hand-held Geiger counter, but there is no similar device for chemical or biological weapons. This makes detecting them by any current technique extremely difficult, since most equipment relies on automating a single test. "*Bomb sniffers*" and the detection equipment used at airports will catch most common types of explosives by testing for nitrates, but most chemical and biological weapons will get right through.

Today, the best approach for determining if a chemical weapon is present involves a technique called *spectroscopy*. Spectroscopy uses a sample's absorption scattering, and other interactions with particular frequencies of light to develop a chemical fingerprint. This is the same approach that astronomers use to determine the atmospheric composition of planets and stars. Unfortunately, it can require a reasonable amount of laboratory equipment and a good quantity of a sample (or a large-scale telescope in the case of stellar measurements).

Results can sometimes take hours to obtain. Biological agents are no easier to detect. When *anthrax-tainted letters* were sent around the country in 2002, it took three full days to give the all-clear at most suspected sites. This kind of time lapse is unacceptable for troops in the field. If an area may be contaminated or a shell may contain chemical or biological agents soldiers on the front lines will have been exposed before they can react, and troops coming up behind them have only minutes to prepare.

In Iraq, the army used pigeons and chickens as an early *chemical weapon* warning system in what some servicemen jokingly called Operation KFC (for *Kuwaiti Fried Chicken*). The idea is that these birds are more vulnerable to the effects of poison gas than people are. If they die, soldiers can conclude that there is a poison in the atmosphere. This was also the reason that miners took canaries into coal mines in the 19th Century. While better than nothing the problems with this scheme are manifold. For a start, almost anything from blowing sand to heatstroke can kill a chicken, so there will be plenty of false alarms. More importantly when much less than a gram of a toxin is enough to kill a persona by the time the chicken dies it is already too late.

Even assuming it were possible to detect chemical and biological toxins instantly and accurately, how could troops be protected from them? Currently, it is necessary to keep troops inside protected vehicles or to make them wear heavy protective Suits that cover the entire body and prevent anything from getting in. This is similar to trying to fight in a space suit, and it isn't practical for active combat or for any mission requiring more than a few hours exposure in desert or jungle conditions.

After the dust settles and the toxins have been detected, what options are there for destroying them arid decontaminating an area? If the toxin is contained within a shell it is relatively easy to destroy, but once it is distributed, it is much more difficult. Some equipment can be power-scrubbed or -cleaned, but the process is expensive, time consuming, and results in a great deal of contaminated water. What if the contamination is in the desert where water is scarce? What if the water supply itself is contaminated? While some biological toxins can be neutralized by boiling or removed by micro-filtration, it is often hard to explain this process to local people in hostile territory, and many chemicals cannot be removed this way. Once they are infected, people can also transmit anthrax and other diseases used as weapons.

How can people be tested quickly and quarantined before they infect others?

These scenarios are more likely than they seem. Saddam Hussein used chemical weapons on Kurds in his own country. He also used eco-terrorist tactics in the Gulf War by setting fire to oil wells. He might well have used chemical weapons to slash and burn his territory had he remained in power long enough after the invasion began. It is clear that if chemical or biological weapons had come into play in Iraq or do come into play before adequate countermeasures are developed, shock and awe military tactics will not help make the U.S. military any less vulnerable. Something much more defensive and more powerful will be needed.

NANOSPHERE

Enter nanotechnology. What if the desk-sized labs that could determine the presence of a chemical or biological toxin could be integrated onto a chip the size of a fingernail? What if they could also perform their tests in seconds or even fractions of a second for the more common toxins? This would make them lightweight enough for any soldier to carry, and quick enough to allow that soldier to take action before it is too late. This may be possible using a nano technology technique called *lab-on-a-chip*. Labs-on-chips are complete chemical or biological laboratories including chemical reactor cells, sample storage, and micro- or nanoscale channels, all integrated onto a microchip in much the same way that circuits are.

Companies such as *Nanosphere* and *Microsensor systems* have already developed the first few devices utilizing this approach for biotoxin detection. Right now these sensors are still the size of laptop computers but they are getting smaller rapidly. There is reason to believe that thumbnail-size sensors will be possible since companies including Agilent (formerly part of Hewlett Packard) and Affymetrix have already made such small sensors for simpler kinds of analysis

A sample device from *nanosphere*, a company specializing in DNA-based sensor nanotechnology. The sensor contains nanoscale particles of gold (often called nano dots), and each of the gold nano dots has several short strands of specifically designed DNA attached to it. The DNA attached to the nano dots will chemically bind to a certain target DNA and to no other; for example, it might bind only to the DNA anthrax. If anthrax is present the DNA strands connected to the nano dots will bind to it, drawing the nanodots closer together. As they draw closer together, the *nano dots* undergo a change that only happens at the nanoscale—they change colour without changing structure or composition.

The reason for the colour change is complex and has to do with the coupling of the properties of quantum mechanics with the properties of the visible world, but the result is straightforward. In the presence of anthrax (or whatever *biotoxin* the sensor is designed to detect) and under no other conditions, the sensor will change colour—in this case, from red to blue. The nano scale colour change may be detected by a laser and transmitted to a soldier via a computer or heads-up display. Alternatively, if a more direct reading is desired, the soldier could carry a visible quantity of these sensors, effectively a litmus paper for biotoxins. While the science behind it may be complicated, this is a test that anyone can use.

This technology and several other promising candidates are evolving quickly, and soon all that will be needed to develop a test for a new toxin is the toxin's genetic fingerprint. The same technology can also be used for diagnosing diseases. In the case of SARS, a genetic fingerprint was found within three months. As DNA sequencing technology continues to develop, this discovery period will get shorter and shorter. With a library of biotoxin fingerprints, a sensor for each toxin could be integrated into a lab-on-a-chip and a universal, nearly instantaneous, very sensitive, and essentially 100% accurate biotoxin sensor could be a reality.

Once a sensor the size of (or very much smaller than) a thumbnail is developed, it could be installed anywhere on a soldier's uniform. Some toxins are very volatile (they evaporate easily) and some are very stable. If a sensor is located on a helmet, for example, it would not be very helpful in detecting *botulinus toxin* in a puddle, whereas a sensor in a boot might not be as helpful in detecting airborne toxins. Ideally, dozens or even hundreds of sensors might be woven into a uniform and interconnected by tiny wires running throughout the fabric. This kind of matrix could still remain extremely light weight and could also serve as a general communications mesh for all of the soldier's equipment, allowing many applications that we will discuss later. A primitive version of this kind of wire mesh is currently used in fencers' jackets to detect the impact of an opponent's sword during competitions.

An alternative to weaving sensors into a uniform is to move the sensors around so they are exposed to everything a soldier is exposed to. One plan for doing this involves injecting sensors into a soldier's bloodstream. The sensors would circulate through the bloodstream and could be monitored at a place where blood vessels are closest to the surface, such as in the eye. The monitoring unit could be installed in goggles or on a microphone boom of a communications headset. While quite invasive, these so-called in vivo sensors could also have other uses in continuously monitoring the health of a soldier.

Since soldiers on the battlefield are already linked by radio communications, any sensor solution should also be tied into the link, perhaps on a separate data channel. If each sensor on every soldier could intercommunicate, a whole unit could start reacting immediately to the presence of a threat as soon as a single sensor was triggered. Also, statistical analysis could help weed out any false positives, making a unit's awareness of threats as close to perfect as possible.

Artificial Materials

Once a *toxin* has been recognized, the next thing to do is to protect the soldier. Since some types of nanotechnology-based sensors can be accurate to the level of detecting individual molecules 0 a toxin, there should be some advance warning before enough toxin to cause fatality has been delivered. How ever, the warning could come only seconds before a lethal dose, far too short a time to put on a chemical defense suit—especially in the middle of a firefight. Fortunately, nanotechnology provides an answer for this, too.

Consider using nanotechnology-enabled smart materials for other purposes. Materials already exist that change their properties when exposed to different external stimuli. Human skin is a good example. It tans in the presence of direct sunlight, sweats when it gets too hot, develops calluses where it is frequently abraded, and heals itself when it is punctured. For these reasons, nanotechnology is of interest to Estee Lauder, Revlon, and others whose customers don't usually dress in combat fatigues. *Artificial materials* with skin like properties have been created, but it is necessary to engineer at the molecular level to create what is required for chemical defense. The idea is to create a fabric that is a comfortable open-weave combat uniform under most conditions but, when exposed to a certain stimulus (an electrical signal from a sensor, for example), could restructure itself at the nanoscale to become an air tight shell. Air could be delivered via an oxygen re-breather of the type used by some divers or through nanotechnology improved gas masks.

Such a uniform is not only possible, but likely. Nanotechnology-enhanced fabrics that are totally resistant to the penetration of liquids (and are thus stain resistant) are already made by companies such as *Nano-Tex*, whose fabric is used in khaki pants sold by Eddie Bauer and others. While you might expect rubber or nylon pants to reject water and stains in this way (although why you would wear rubber pants is another question), only nanotechnology can combine the properties of comfortable cotton fibers with the desirable properties of repelling liquid to make fabric that is indistinguishable from ordinary khakis until you spill on it. Employ so-called *smart material* in soldiers' uniforms is just an evolution of the techniques already in use. As we'll see later, that is only one of the ways nanotechnology will affect soldier's' uniforms.

It is important to protect a soldier in his uniform, but better defenses are also required for buildings, vehicles, and other installations. Filtering out biotoxins is in some ways simpler than filtering out chemical weapons, since a single virus or bacterium is more than 100 times larger than a small molecule such as HCN or oxygen. This means that a filter with pores measuring tens of nanometers can be made to block out most biotoxins but to let air through. Unfortunately, these filters don't work well for chemical weapons where the difference in size between air and toxin may be very small indeed.

Right now, most chemical weapons protection is based on a similar technology to World I—*era gas masks*. A simple air filter is impregnated with activated carbon (also used in commercially available water filters) and other chemicals that adsorb many common contaminants. This approach is far from effective in many cases. For example, HCN is one of many common chemical weapons that does not react much with activated carbon.

A different approach involves making an artificial membrane and permeating it with nanotubes. *Nanotubes* are shaped like drinking straws made of carbon, and their diameters can be controlled with sub-nanometer precision. Using this approach could create a molecular sieve that would block all but the very smallest molecules—oxygen and nitrogen, for example and make a very good filter indeed. This effect depends on tailoring the sizes of individual molecules and is only possible through nanotechnology.

Cleaning the Contamination

If an area is contaminated with chemical or biological agents, it must be cleaned up in a process commonly called *remediation*. Remediation can be done in a variety of ways, including old-fashioned soaking with soap and water, forcing breakdown via chemical reaction, or exposing to extreme heat. Nano science offers a few ways to vastly improve the efficiency of this process. Consider the remediation approach of breaking down a toxic substance through a chemical reaction. A material's surface area and its reactivity are directly related. The greater the surface area is, the greater is the exposure to other materials with which to bond and interact. This is one way in which catalysts (materials that increase the speed of chemical reactions) work. Grinding a substance finer and finer increases surface area and the nanoscale is the logical limit. A quarter ounce of nano particles may have a surface area as great as a football field.

Some powders, such as *aluminum oxide* and *magnesium oxide*, have proven effective in neutralizing chemical agents and are already in use by the army. Reducing these particles to the nano scale not only increases their effectiveness by increasing reactivity, but also allows them to be impregnated into the fibers of filters like the gas mask filters we just discussed. Effectively this allows them to do what activated carbon did in older filters—aggressively bind to and remove chemical agents. While activated carbon does work this way, it is only one example of a whole family of nano particles that can be developed to deal with a host of chemical weapons. Activated carbon was created by trial and error rather than through a knowledge of the underlying mechanism that makes it function. However, research in nano science is revealing these mechanisms and is allowing the design of new nano particles specifically engineered to react with many chemicals that activated carbon can't handle.

The examples above are some of the ways in which nanotechnology will protect soldiers from *chemical* and *biological weapons*. Sensors will instantly detect the threat, transmit it to a central control unit, and initiate countermeasures. The soldier's uniform will immediately start to defend against attack by becoming nonporous and filtering the air supply. *Nanoparticles* can then be sprayed, released, or even delivered via artillery shell or other munitions to start breaking down the *chemical* or *biological agent*. Clearly these technologies have other dual-use applications including medical diagnosis using DNA-based sensors and water purification using remediation nano particles. There are also a number of other nanotechnologies that may be effective in countering weapons of mass destruction, but this is a great beginning.

Battlefield Computers

In ancient Greece, a soldier went into battle carrying over 70 pounds of weapons and equipment, including heavy bronze armor, spears, a shield, and a sword. A modern soldier carries at least as much, though the nature of what he carries has changed significantly. Today's soldiers are expected to carry not just weapons and body armor, but radio equipment, optics such as night-vision or telescopic scopes, *global positioning systems* (GPSs), *battlefield computers*, food, water, spare clothing, and much, much more—the batteries alone can weigh more than 20 pounds. These tools are essential for net worked warfare and battle space awareness. It is crucial for troops to gather information, coordinate, maneuver, respond to enemy action, and remain in touch with commanders even in the heat of battle.

While these tools are effective, they also have serious problems. Seventy to ninety pounds of equipment are unwieldy and lead to fatigue. The systems are not integrated with each other—most hand-held range finders cannot automatically transmit information to weapons systems, and binoculars and scopes don't transmit image information, Geiger counters don't directly relay danger information to other units, and most devices can't even share power supplies so they require heavy battery units or other portable power supplies. While many devices provide a soldier with information, the devices do not process it or respond to it automatically. *Ballistic defensive* technologies (such as Kevlar vests) offer no protection for the arms, legs, or head, and they don't monitor vital signs or take action if a soldier is wounded. In many respects modern soldiers are equipped much like Spartan warriors, but with a better spear, tougher armor, and a lot of electronics. This bundle of equipment has caused them to be compared to a Christmas tree with ornaments hanging from it. The green uniform probably does not help either.

The soldier of the future could look very different. A great deal of effort is going into making soldiers more efficient and increasing survivability, a soldier's ability to avoid injury or recover if he or she is injured. The U.S. Army is very serious about this effort. It has already spent $50 million creating a Center for Soldier Nanotechnologies at MIT, and it is investing a great deal of money in other research efforts as well.

Camouflage

Nanotechnology will directly affect the war fighter at all levels. One good example of what the future may hold is the uniform. *Combat fatigues* or *BDUs* are one of the most overlooked parts of a soldier's kit. They are dyed with camouflage patterns and the fabric is resilient, but there is little else to recommend them. When we discussed weapons of mass destruction, we introduced the idea of making the uniform into a matrix to which sensors could be attached. This is only the beginning.

Protecting a soldier is important, but ideally he or she should also be difficult to see and therefore difficult to attack. This is the idea behind current camouflage uniforms, stealth bombers, and the ability to fight effectively at night—advanced night vision gear is already based on nanotechnology and represents an early application.

The *camouflage* in current uniforms is called *passive camouflage*, meaning that it doesn't change or adapt to different circumstances. *Active camouflage*, however, is just the opposite—it blends in with whatever environment it is exposed to, just as a *chameleon blends* in with rocks, leaves, or soil. A chameleon like approach to making soldiers very difficult to detect may be possible through nanotechnology and *biomimicry*. By working at the same scale as nature, natural processes like those of the chameleon can be emulated and integrated into useful devices. James Bond fans recently saw an exaggerated version of active camouflage in *Die Another Day* in which 007's car could disappear into the ice with the flick of a switch. While active camouflage probably won't be good enough to fool someone up close, active camouflage could certainly have applications for jets and aircraft, making them as hard to see with the eye as they are hard to see with radar.

While it is best to avoid injury through stealth wherever possible, it is necessary to minimize the damage a soldier sustains in an attack should one occur. One approach to doing this involves improving Kevlar the fabric used to make most bulletproof vests. Kevlar is a polymer whose shape is rigid as opposed to the more pliable, thread-like structure of most polymer chains. This rigidity allows Kevlar to disperse energy from local impacts pretty well and to prevent bullets from getting through. Very new smart materials being developed by Ray Baughman's group at the University of Texas in Dallas incoporate nanotube fibers into an open-weave cloth that is more than four times tougher than spider silk and seventeen times tougher than Kevlar.

Researchers at MIT are pursuing another approach. This involves exploring the potential of making a uniform that not only reacts to chemical or biological toxins, but can stiffen and act as armor against ballistic threats such as bullets and fragmentation. Imagine a sheet of paper held at one end. It is floppy and flexible, but if you curve it slightly as you hold it, it becomes stiff and rigid. One approach to allowing a uniform to act as armor involves packing nano particles of iron onto the fabric's fibers. Normally these particles do little other than add a bit of weight, but in the presence of a magnetic field they align and force the fibers to become rigid. The rigid fibers could offer significant protection against ballistic threats and the armor could be activated or deactivated for convenience, depending on whether the soldier is in battle or just patrolling. While iron nano particles may add a bit of weight, they will be significantly lighter than the 15-pound vests that are common today.

Unfortunately, while improved armor will reduce injuries, even nano-protected soldiers can still be hurt in battle. In recent conflicts, some 50% of battlefield fatalities were not immediate but were the result of bleeding or complications occurring several minutes, hours, or even days after a wound was inflicted. Nanotechnology may help to reduce this problem.

If a soldier is hurt in the middle of a firefight, he or she must wait for some kind of assistance from a corpsman or a comrade with a medical kit. What if the nano-enhanced uniform could help with this, too? Sensors could be woven into the fabric that could look for hemoglobin, the oxygen-carrying compound in the blood. Good hemoglobin sensors have already been developed for medical diagnostics such as the detection of blood in stool samples, and these could be incorporated into labs-on-chips and used to begin treatment. Along with the hemoglobin sensors, the tiny labs could also contain supplies of antiseptics, antibiotics, and anesthetics. The fabric itself could contain a smart material liner that would adhere to the wound and act as a temporary bandage.

Groups like the Institute for New Materials in Germany have already demonstrated using *biocidal nanoparticles* (particles that kill bacteria on contact) to keep medical equipment such as hearing aids sterile. These nano particles aren't made of anything exotic—silver is one of the best nano-biocides. It has been suggested that these kinds of biocidal particles could be incorporated into hospital bed sheets to reduce the incidence of cross-infection, and they could be used in uniforms too. Wearing a biocidal uniform could help reduce infection from injuries as well as reduce other kinds of contamination that can occur when uniforms cannot be changed or cleaned regularly. It is not uncommon for a soldier in the battlefield to go for several days working hard, sweating and perhaps bleeding, without having an opportunity to change clothes or visit the laundry.

We began this section with a criticism of a *soldier's kit*. We described it as too cumbersome, and too bulky. We accused it of a lack of integration. Yet now we are suggesting adding a wire mesh, hundreds of sensors, iron nano particles for protection, silver nanoparticles for the reduction of infection, nanoscale medical kits, and who knows what to enable *active camouflage*. It sounds like we've made the situation rather worse than better, but this is not the case.

First of all, these technologies are not as unrelated as they seem. They should all be integrated into a single, coherent system—the next generation uniform. This system will contain a central communications channel in the form of a wire mesh or similar solution and integrated processors to handle information. It will contain functional components or subsystems that react to specific threats. Many of these subsystems have multiple applications. A sensor can be used to detect an external threat, but it may also be used to detect blood and apply treatment. Clothing that can change from flexible to stiff may function as armor or as a cast or tourniquet for medical purposes. One approach to integrating all of these solutions into a uniform involves placing nanoscale channels into the core of the fibers within the fabric. Not only would this solution enable the applications we've discussed, but the electromechanical channel opens the door to using the uniform to actually enhance human performance. It might, for example, augment human muscles and make the wearer effectively stronger, thus providing a unique solution to the weight problem.

Whether they enhance performance or not, weight may not be an issue for such a uniform. To understand why, remember just how small the nanoscale is. Ask anyone who has worn a pair of Nano Care pants if there is any difference in weight between them and ordinary khakis. Ask anyone who has used a Wilson Double-Core tennis ball (based on nanotechnology enhanced composites) if there is any difference in weight between them and ordinary tennis balls. See if you can feel the difference in weight between a microchip with copper technology and one based on aluminum. While it isn't certain, it is likely that one or more of the improvements we have talked about will be possible without affecting the weight or comfort a uniform except in a positive direction.

Ground forces tend to remain in trouble spots long after the Air Force, Navy, and Marines have pulled out. They act as police, peacekeepers and even engineers in the restoration of basic services after fighting is over. They need a better layer of lightweight protection to help them get these jobs done. Nanotechnology provides the only likely solutions to these problems.

Superhuman

The tasks of modern soldiers might well be called *superhuman* and thus require superhuman characteristics to accomplish them. We briefly mentioned using channels in the fibers of uniforms to provide performance enhancement for soldiers, but a lot of other approaches are possible as well.

One performance enhancement possibility for increasing strength and reflexes, as well as flying aircraft or driving cars, is a direct interface between the human nervous system and electronics. There have been many promising results in this direction, particularly in the area of prosthetics and bionics. Hearing aids, for example used to work by simply amplifying the sounds around them. *Neuro-electronic hearing aids* work by actually transforming (technically the term is transducing) the sound into a neural impulse and injecting it into the nervous system bypassing the eardrum entirely. Similarly, neuro-electronic eyes attempt to project images directly onto the retina. These devices were created to solve the most untreatable forms of blindness and deafness and are having good results. Optobionics, a company that develops vision solutions, has demonstrated that some trial patients who could see only darkness now see blurry shapes and those who saw blurry shapes can now distinguish between teams at a ball game.

This kind of interface promises treatments for all sorts of conditions from missing limbs to quadriplegia. As it becomes less invasive it could be used to eliminate the old-fashioned steering yoke so pilots can fly planes directly (giving the expression "*fly-by-wire*" a whole new meaning) or to control servo-motors in a battle suit which could increase speed, strength, protection, and more. Working at the scale of nature will make this possible.

Unmanned Vehicle

Perhaps one of the most controversial ideas for protecting soldiers involves taking them out of combat. This does not mean that wars and conflicts can be entirely avoided, but it may be possible to use more and more mechanization to fight future wars.

Consider the Predator UAV (*Unmanned Aerial Vehicle*). The Predator is just one in a family of unmanned aircraft produced by General Atomics, but in 2001 it made history as the first unmanned aircraft to launch a live antitank missile in combat. While not strictly based on nanotechnology, the Predator shows an interesting trend. Remotely controlled combat aircraft are now a reality as are autonomous missile systems like the Tomahawk cruise missile, which can find a target using GPS and other programming even over complex terrain. Indeed, missiles guided by lasers and heat are now old hat.

Ground vehicles, however are not yet automated. Unlike Predator or Tomahawk, the missions of unmanned ground vehicles do not involve simply going to a location, attempting a kill, and, in the case of Predator, returning to base. A tank or other ground combat vehicle must negotiate changing terrain and deal with rapidly changing threats. Even the short period of latency (the time between when a signal is sent and received) between it and a remote controller could make a world of difference. This is very similar to the set of challenges NASA faces in creating landing vehicles for Mars. To be successful, it will have to be somewhat autonomous, and it will likely rely on several forms of information technology and processing that are not yet fully developed. These include image recognition and, more controversially, artificial intelligence.

It may be a long time before the military or the public is ready for unmanned vehicles or even robots accompanying soldiers into battle, but DARPA (the *Defense Advanced Research Projects Agency*) and others have been looking at the problem for some years.

Several of the components required to make autonomous land vehicles will come from nanotechnology. These include high-performance computers using new architectures that may make artificial intelligence a reality. This artificial intelligence will not necessarily resemble human intelligence, but it will more likely be a specialized system capable of making intelligent decisions of a certain type, a sort of expert system. Thus far even the most advanced silicon computers have been able to show only modest results in this regard since they are designed on an architecture optimized for numerical computation, not cognitive intelligence. Nanotechnological approaches to computing including the use of molecular electronics, bioelectronics (electronic components integrating proteins and other biological entities), and quantum computing may change this.

Speedy Robots

While autonomous robotic land vehicles may still be in the future, nanotechnology has changes in store for conventional combat vehicles, such as fighters, helicopters, and tanks. Most of these changes will come from the application of nanotechnology-based smart materials.

For example, consider stealth aircraft. While the most advanced models are said to have radar profiles comparable to that of a bumblebee, they make large sacrifices in terms of design. The first stealth fighter, the F-117, has an exotic, panel-shaped design and is far from being the most maneuverable or speedy fighter in the sky. The F/A-22 and other later stealth designs have improved on the F-117, and nanotechnology will help still more. Nanoparticles with massive absorption cross sections for radar and infrared are being developed. This means that they soak tip a huge amount of energy and can therefore prevent detection. In addition, the same active camouflage we discussed for soldiers' uniforms may be applied to aircraft and ground vehicles.

Nanotechnology will also make vehicles faster. Argonide Corporation already produces *Alex*, a rocket-fuel additive based on nano particles of aluminum. *Alex* vastly increases the efficiency of burning hydrocarbon fuels like kerosene and 0 have applications for jet fuels. Such an additive could also increase the efficiency of powder burn in a rifle cartridge.

Another barrier to speed, fuel efficiency, and deployment is the weight of combat vehicles. A tank may weigh 60 tons and infantry vehicles weigh almost as much. Lighter materials such as the ever-

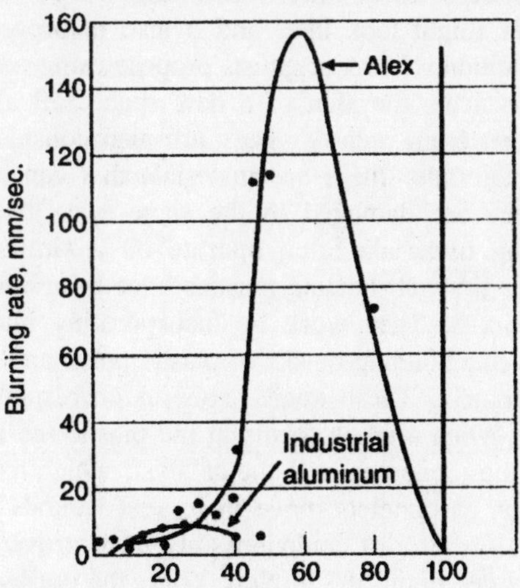

Fig. 8.1. This chart shows the relative performance of Alex, a nanopowdered fuel additive, and its microparticle competitors.

useful carbon nano tube may help solve this problem. Next-generation commercial aircraft already utilize car bon fiber and other super strong, lightweight carbon composites in their construction, but none of these materials can come close to the prowess of nanotubes. With strength on the order of 60 times that of steel at a fraction of the weight, nanotubes will be used to make aircraft that are faster and that require shorter runways and less fuel. Nanotubes or other such nanomaterials can even be used to armor aircraft such as attack helicopters which are usually lightly armored for weight reasons. Companies like CNI and NanoLab are now offering nanotubes in appreciable quantities, but they still can cost about a hundred dollars a gram. Still, this price is far lower than it was just two years ago, and two years before that the idea of bulk nano tubes was still a pipe dream.

While *carbon nanotubes* may not yet be the armor of choice, nano-enhanced armor is already in use protecting troops. The SAG (Save a Gunner) turret, for light combat vehicles such as HMMWVs is a case in point. Produced by US Global Nanospace, it uses nanofiber materials to create a gun turret that is less than 200 pounds, has minimal impact on mobility, and yet can withstand six successive 7.62-caliber bullet strikes in the same spot. US Global Nanospace uses similar technology to make blast-proof doors for aircraft and materials for dealing with explosives.

NASA engineers have started thinking about what its next-generation orbiter might look like, and it also incorporates many of these ideas. In addition, NASA scientists propose using nanotechnology-based coatings to make the skin of a new spacecraft almost friction free, reducing heat from reentry. They are also considering the idea of self-healing materials—these are materials that can start to repair themselves if they are damaged in the same way that human skin does. Self-healing materials often operate on a similar premise to skin. Some of the first self-healing plastics have been developed at the University of Illinois. They work by incorporating microspheres of liquid monomers (the building blocks of plastic polymers) and a catalyst into the plastic matrix. These micro- spheres correspond to platelets in human blood. When a crack forms in the plastic the micro spheres break to expose the monomers and the catalyst, which then move in to seal the crack just as platelets move in to seal wounds in skin. Self-healing materials on aircraft could possibly have prevented the crash of American Airlines flight 587 in New York, the crash of Air France Concorde flight 4590 in Paris, and even the explosion of the space shuttle Columbia. Aside from their applications for civil aviation, they could also make combat aircraft much more survivable.

ALTERNATIVE ENERGY SOURCES

Alternative energy sources will be necessary to take advantage of many of these new nanotechnologies. Current equipment is already power-hungry, and many of the technologies we have discussed will exacerbate the problem. Soldiers currently carry more than 20 pounds of batteries and portable power supplies into battle, but consider the issues of other kinds of combat equipment. Ground vehicles and aircraft are powered using hydrocarbon fuels (usually diesel or kerosene, although some multi-fuel engines can also run on gasoline or even alcohol). These fuel supplies are very volatile and can result in "secondary explosions," blasts that occur after an attack usually as a result of hazardous cargo such as fuel or ammunition. Fuel depots represent huge targets for the enemy and transporting fuel is a logistical nightmare.

Naval vessels have dealt with this problem in many cases by using nuclear fission reactors. These have the advantage that they seldom need to be refueled (only every few years) and thus give the vessels effectively unlimited operational range, but they also create other hazards. If a nuclear vessel is disabled, the fissionable material may be recovered and used for making a bomb. If it is destroyed or damaged the radioactive materials it contains could cause an environmental catastrophe.

Promising nanotechnology-based energy solutions such as fuel cells, efficient solar power, and even nuclear fusion (which has a nano component) could reduce or eliminate these concerns.

NANOARCHITECTURE

The applications we have discussed in this chapter represent only the beginning of the possible applications of nanotechnology for defense. Nanotechnology will greatly increase the survivability of soldiers and equipment, especially for peacekeepers in the period after the hottest fighting is over and when a full combat rig is impractical. It will provide stealthier, tougher and faster combat vehicles and increase battle space information and awareness. It will improve communication within Units and allow a soldier's kit to become an efficient, integrated system. It will combat the threats of chemical an biological weapons. It will be a crucial tool for weapon inspectors and those looking for weapons of mass destruction. It will provide lighter; more efficient portable power. It already helps our troops to see at night and provides some of the few chemical and biological defenses that we have so far developed. For these reasons, nanotechnology is key to the military's new mission.

When he developed one of the first machine guns during the Civil War, Richard Gatling sincerely hoped that the destruction it would wreak would be so great that no one would dare to use it and wars would become impossible. Clearly this did not happen, and the advice that guided one of his successors, Hiram Maxim, seemed much more apt: "If you wanted to make a lot of money, invent something that will enable these Europeans to cut each other's throats with greater facility." Perhaps Gatling's vision was most nearly attained by the nuclear weapon, but even the MAD doctrine depends on sane and stable governments having exclusive control of weapons of mass destruction There are no credible claims that nanotechnology will take the final step in making war obsolete, hut one thing that makes nanotechnology particularly exceptional among technologies with military applications is that it is primarily useful for defense. It can enhance soldier safety and capabilities, limit the destruction necessary to accomplish objectives, provide information and intelligence, coordinate troops, enhance reaction times, and even help heal the wounded. It may also make a lot of money. While it certainly could be used for weapons, nuclear weapons are still the ultimate killing machine. Since the American military already has destructive weapons of almost limitless capability, what it needs now is better defense and better capability for using its offensive weapons effectively. That is where nanotechnology will start.

9

WOUND HEALING

Since the human body is made of molecules, molecular technology will be used to bring health. The ill, the old, and the injured all suffer from misarranged patterns of atoms, whether misarranged by invading viruses, passing time, or swerving cars. Devices able to rearrange atoms will be able to set them right. *Nanotechnology* will bring a fundamental breakthrough in medicine. Physicians now rely chiefly on surgery and drugs to treat illness. Surgeons have advanced from stitching wounds and amputating limbs to repairing hearts and reattaching limbs. Using microscopes and fine tools, they join delicate blood vessels and nerves. Yet even the best microsurgeon cannot cut and stitch finer tissue structures. Modern scalpels and sutures are simply too coarse for repairing capillaries, cells, and molecules. Consider "delicate" surgery from a cell's perspective a huge blade sweeps down, chopping blindly past and through the molecular machinery of a crowd of cells, slaughtering thousands. Later, a great obelisk plunges through the divided crowd, dragging a cable as wide as a freight train behind it to rope the crowd together again. From a cell's perspective even the most delicate surgery, performed with exquisite knives and great skill, is still a butcher job. Only the ability of cells to abandon their dead, regroup, and multiply makes healing possible. Yet as many paralyzed accident victims know too well, not all tissues heal.

Drug therapy, unlike surgery, deals with the finest structures in cells. *Drug molecules* are simple molecular devices. Many affect specific molecules in cells. *Morphine molecules*, for example, bind to certain receptor molecules in brain cells, affecting the neural impulses that signal pain. *Insulin*, *beta blockers*, and other drugs fit other receptors. But drug molecules work without direction. Once dumped into the body, they tumble and bump around in solution haphazardly until they bump a target molecule, fit, and stick, affecting its function. Surgeons can see problems and plan actions, but they wield crude tools; drug molecules affect tissues at the molecular level, but they are too simple to sense, plan, and act. But molecular machines directed by nano computers will offer physicians another choice. They will combine sensors, programs, and molecular tools to form systems able to examine and repair the ultimate components of individual cells. They will bring surgical control to the molecular domain.

These advanced molecular devices will be years in arriving, but researchers motivated by medical needs are already studying molecular machines and molecular engineering. The best drugs affect specific molecular machines in specific ways. Penicillin, for example, kills certain bacteria by jamming the *nanomachinery* they use to build their cell walls, yet it has little effect on human cells.

Biochemists study molecular machines both to learn how to build them and to learn how to wreck them. Around the world (and especially the Third World) a disgusting variety of viruses, bacteria, protozoa, fungi, and worms parasitize human flesh. Like penicillin, safe, effective drugs for these diseases would jam the parasite's molecular machinery while leaving human molecular machinery unharmed. Dr. Seymour Cohen, professor of pharmacological science at Suny, argues that biochemists should systematically study the molecular machinery of these parasites. Once biochemists have determined the shape and function of a vital protein machine, they then could often design a molecule shaped to jam it and ruin it. Such drugs could free humanity from such ancient horrors as schistosomiasis and leprosy, and from new ones such as AIDS.

Drug companies are already redesigning molecules based on knowledge of how they work. Researchers at Upjohn Company have designed and made modified molecules of *vasopressin*, a hormone that consists of a short chain of amino acids. Vasopressin increases the work done by the heart and decreases the rate at which the kidneys produce urine; this increases blood pressure. The researchers de signed modified vasopressin molecules that affected receptor molecules in the kidney more than those in the heart, giving them more specific and controllable medical effects. More recently, they de signed a modified vasopressin molecule that binds to the kidney's receptor molecules without direct effect, thus blocking and *inhibiting* the action of natural vasopressin.

Medical needs will push this work forward, encouraging researchers to take further steps toward protein design and molecular engineering. Medical, military, and economic pressures all push us in the same direction. Even before the assembler breakthrough, *molecular technology* will bring impressive advances in medicine; trends in biotechnology guarantee it. Still, these advances will generally be piecemeal and hard to predict, each exploiting some detail of biochemistry. Later, when we apply assemblers and technical AI systems to medicine, we will gain broader abilities that are easier to foresee.

To understand these abilities, consider cells and their self-repair mechanisms. In the cells of your body, natural radiation and noxious chemicals split molecules, producing reactive molecular fragments. These can misbond to other molecules in a process called cross-linking. As bullets and blobs of glue would damage a machine, so radiation and reactive fragments damage cells, both breaking molecular machines and gumming them up. If your cells could not repair themselves, damage would rapidly kill them or make them run amok by damaging their control systems. But evolution has favoured organisms with machinery able to do something about this problem. The self-replicating factory system repaired itself by replacing damaged parts; cells do the same. So long as a cell's DNA remains intact, it can make error-free tapes that direct ribosomes to assemble new protein machines.

Unfortunately for us, DNA itself becomes damaged, resulting in mutations. Repair enzymes compensate somewhat by detecting and repairing certain kinds of damage to DNA. These repairs help cells survive, but existing repair mechanisms are too simple to correct all problems, either in DNA or elsewhere. Errors mount, contributing to the aging and death of cells—and of people.

Life, Mind, and Machines

Does it make sense to describe cells as "machinery," whether self- repairing or not? Since we are made of cells, this might seem to reduce human beings to "*mere machines*," conflicting with a holistic understanding of life.

But a dictionary definition of holism is "the theory that reality is made up of organic or unified wholes that are greater than the simple sum of their parts." This certainly applies to people: one simpler sum of our parts would resemble hamburger, lacking both mind and life. The human body includes some ten thousand billion *billion* protein parts, and no machine so complex deserves the label "mere." Any brief description of so complex a system cannot avoid being grossly incomplete, yet at the cellular level a description in terms of machinery makes sense. Molecules have simple moving parts, and many act like familiar types of machinery. Cells considered as a whole may seem less mechanical, yet biologists find it useful to describe them in terms of molecular machinery.

Biochemists have unraveled what were once the central mysteries of life, and have begun to fill in the details. They have traced how molecular machines break food molecules into their building blocks and then reassemble these parts to build and renew tissue. Many details of the structure of human cells remain unknown (single cells have billions of large molecules of thousands of different kinds), but biochemists have mapped every part of some viruses. Biochemical laboratories often sport a large wall chart showing how the chief molecular building blocks flow through bacteria. Biochemists under stand much of the process of life in detail, and what they don't understand seems to operate on the same principles. The mystery of heredity has become the industry of genetic engineering. Even embryonic development and memory are being explained in terms of changes in biochemistry and cell structure.

In recent decades, the very quality of our remaining ignorance has changed. Once, biologists looked at the process of life and asked, "How can this be?" But today they understand the general principles of life, and when they study a specific living process they commonly ask, "Of the many ways this could be, which has nature chosen?" In many instances their studies have narrowed the competing explanations to a field of one. Certain biological processes—the coordination of cells to form growing embryos, learning brains, and reacting immune systems-still present a real challenge to the imagination. Yet this is not because of some deep mystery about how their parts work, but because of the immense complexity of how their many parts interact to form a whole.

Cells obey the same natural laws that describe the rest of the world. *Protein machines* in the right molecular environment will work whether they remain in a functioning cell or whether the rest of the cell was ground up and washed away days before. Molecular machines know nothing of "*life*" and "*death*."

Biologists—when they bother—sometimes define life as the ability to grow, replicate, and respond to stimuli. But by this standard, a mindless system of replicating factories might qualify as life, while a conscious artificial intelligence modeled on the human brain might not. Are viruses alive, or are they "*merely*" fancy molecular machines? No experiment can tell, because nature draws no line between living and nonliving. Biologists who work with viruses instead ask about viability: "Will this virus function, if given a chance?" The labels of "*life*" and "*death*" in medicine depend on medical capabilities: physicians ask, "Will this patient function, if we do our best?" Physicians once declared patients dead when the heart stopped; they now declare patients dead when they despair of restoring brain activity. Advances in cardiac medicine changed the definition once; advances in brain medicine will change it again.

Just as some people feel uncomfortable with the idea of machines thinking, so some feel uncomfortable with the idea that machines underlie our own thinking. The word "*machine*" again seems to conjure up the wrong image, a picture of gross, clanking metal, rather than signals flickering through a shifting weave of neural fibers, through a living tapestry more intricate than the mind it embodies can fully comprehend. The brain's really machinelike machines are of molecular size, smaller than the finest fibers.

A whole need not resemble its parts. A solid lump scarcely resembles a dancing fountain, yet a collection of solid, lumpy molecules forms fluid water. In a similar way, billions of molecular machines make up neural fibers and synapses, thousands of fibers and synapses make up a neural cell, billions of neural cells make up the brain, and the brain itself embodies the fluidity of thought. To say that the mind is "*just molecular machines*" is like saying that the Mona Lisa is '*just dabs of paint*." Such statements confuse the parts with the whole, and confuse matter with the pattern it embodies. We are no less human for being made of molecules.

From Drugs to Cell Repair Machines

Being made of molecules, and having a human concern for our health, we will apply molecular machines to biomedical technology. Biologists already use antibodies to tag proteins, enzymes to cut and splice DNA, and viral syringes (like the T_4 phage) to inject edited DNA into bacteria. In the future, they will use assembler-built nanomachines to probe and modify cells. With tools like disassemblers, biologists will be able to study cell structures in ultimate, molecular detail. They then will catalog the hundreds of thousands of kinds of molecules in the body and map the structure of the hundreds of kinds

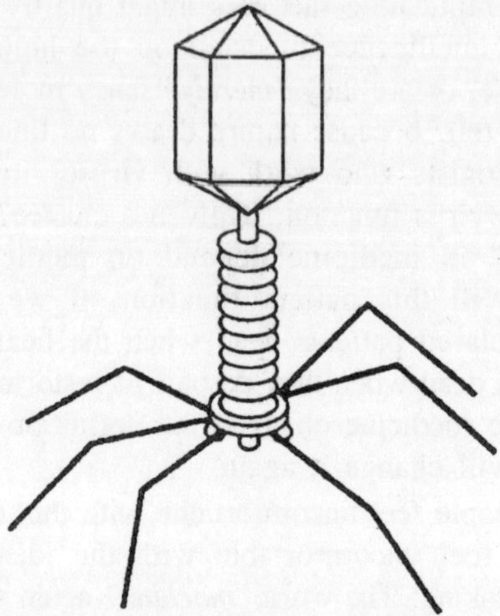

Fig. 9.1. Bacteriophage T_4

of cells. Much as engineers might compile a parts list and make engineering drawings for an automobile, so biologists will describe the parts and structures of healthy tissue. By that time, they will be aided by sophisticated technical AI systems. Physicians aim to make tissues healthy, but with drugs and surgery they can only encourage tissues to repair themselves. Molecular machines will allow more direct repairs, bringing a new era in medicine.

To repair a car, a mechanic first reaches the faulty assembly, then identifies and removes the bad parts, and finally rebuilds or replaces them. Cell repair will involve the same basic tasks—tasks that living systems already prove possible.

Access

White blood cells leave the bloodstream and move through tissue, and viruses enter cells. Biologists even poke needles into cells without killing them. These examples show that molecular machines can reach and enter cells.

Recognition

Antibodies and the tail fibers of the T_4 phage—and indeed, all specific biochemical interactions—show that molecular systems can recognize other molecules by touch.

Disassembly

Digestive enzymes (and other, fiercer chemicals) show that molecular systems can disassemble damaged molecules.

Rebuilding

Replicating cells show that molecular systems can build or rebuild every molecule found in a cell.

Reassembly

Nature also shows that separated molecules can be put back together again. The machinery of the T_4 phage, for example, self-assembles from solution, apparently aided by a single enzyme. Replicating cells show that molecular systems can assemble every system found in a cell.

Thus, nature demonstrates all the basic operations that are needed to perform molecular-level repairs on cells. What is more, systems based on nanomachines will generally be more compact and capable than those found in nature. Natural systems show us only lower bounds to the possible, in cell repair as in everything else.

Cell Repair Machines

In short, with molecular technology and technical AI we will compile complete, molecular-level descriptions of healthy tissue, and we will build machines able to enter cells and to sense and modify their structures.

Cell repair machines will be comparable in size to bacteria and viruses, but their more-compact parts will allow them to be more complex. They will travel through tissue as white blood cells do, and enter cells as viruses do—or they could open and close cell membranes with a surgeon's care. Inside a cell, a repair machine will first size up the situation by examining the cell's contents and activity, and then take action. Early cell repair machines will be highly specialized, able to recognize and correct only a single type of molecular disorder, such as an enzyme deficiency or a form of DNA damage. Later machines (but not much later, with advanced technical AI systems doing the design work) will be programmed with more general abilities.

Complex repair machines will need nanocomputers to guide them. A micron-wide mechanical computer will fit in 1/1000 of the volume of a typical cell, yet will hold more information than does the cell's DNA. In a repair system, such computers will direct smaller, simpler computers, which will in turn direct machines to examine, take apart, and rebuild damaged molecular structures.

By working along molecule by molecule and structure by structure, repair machines will be able to repair whole cells. By working along cell by cell and tissue by tissue, they (aided by larger devices, where need be) will be able to repair whole organs. By working through a person organ by organ, they will restore health. Because molecular machines will be able to build molecules and cells from scratch, they will be able to repair even cells damaged to the point of complete inactivity. Thus, cell repair machines will bring a fundamental breakthrough: they will free medicine from reliance on self- repair as the only path to healing.

To visualize an advanced cell repair machine, imagine it—and a cell—enlarged until atoms are the size of small marbles. On this scale, the repair machine's smallest tools have tips about the size of your fingertips; a medium-sized protein, like hemoglobin, is the size of a typewriter; and a ribosome is the size of a washing machine. A single repair device contains a simple computer the size of a small truck, along with many sensors of protein size, several manipulators of ribosome size, and provisions for memory and motive power. A total volume ten meters across, the size of a three-story house, holds all these parts and more. With parts the size of marbles packing this volume, the repair machine can do complex things.

But this repair device does not work alone. It, like its many siblings, is connected to a larger computer by means of mechanical data links the diameter of your arm. On this scale, a cubic-micron computer with a large memory fills a volume thirty stories high and as wide as a football field. The repair devices pass it information, and it passes back general instructions. Objects so large and complex are still small enough: on this scale, the cell itself is a kilometer across, holding one thousand times the volume of a cubic-micron computer, or a million times the volume of a single repair device. Cells are spacious.

Will such machines be able to do everything necessary to repair cells? Existing molecular machines demonstrate the ability to travel through tissue, enter cells, recognize molecular structures, and so forth, but other requirements are also important. Will repair machines work fast enough? If they do, will they waste so much power that the patient will roast?

The most extensive repairs cannot require *vastly* more work than building a cell from scratch. Yet molecular machinery working within a cellular volume routinely does just that, building a new cell in tens of minutes (in bacteria) to a few hours (in mammals). This indicates that repair machinery occupying a few percent of a cell's volume will be able to complete even extensive repairs in a reason able time—

days or weeks at most. Cells can spare this much room. Even brain cells can still function when an inert waste called *lipofuscin* (apparently a product of molecular damage) fills over ten percent of their volume.

Powering repair devices will be easy: cells naturally contain chemicals that power *nanomachinery*. Nature also shows that repair machines can be cooled: the cells in your body rework themselves steadily, and young animals grow swiftly without cooking themselves. Handling heat from a similar level of activity by repair machines will be no sweat—or at least not too much sweat, if a week of sweating is the price of health.

All these comparisons of repair machines to existing biological mechanisms raise the question of whether repair machines will be able to *improve* on nature. DNA repair provides a clear-cut illustration.

Just as an illiterate "*book-repair machine*" could recognize and repair a torn page, so a cell's repair enzymes can recognize and repair breaks and cross-links in DNA. Correcting misspellings (or mutations), though, would require an ability to read. Nature lacks such repair machines, but they will be easy to build. Imagine three identical DNA molecules, each with the same sequence of nucleotides. Now imagine each strand mutated to change a few scattered nucleotides. Each strand still seems normal, taken by itself. Nonetheless, a repair machine could compare each strand to the others, one segment at a time, and could note when a nucleotide failed to match its mates. Changing the odd nucleotide to match the other two will then repair the damage. This method will fail if two strands mutate in the same spot. Imagine that the DNA of three human cells has been heavily damaged— after thousands of mutations, each cell has had one in every million nucleotides changed. The chance of our three-strand correction procedure failing at any given spot is then about one in a million mil lion. But compare five strands at once, and the odds become about one in a million million-million, and so on. A device that compares many strands will make the chance of an uncorrectable error effectively nil.

In practice, repair machines will compare DNA molecules from several cells, make corrected copies, and use these as standards for proofreading and repairing DNA throughout a tissue. By comparing several strands, repair machines will dramatically improve on nature's repair enzymes.

Other repairs will require different information about healthy cells and about how a particular damaged cell differs from the norm. Anti bodies identify proteins by touch, and properly chosen antibodies can generally distinguish any two proteins by their differing shapes and surface properties. Repair machines will identify molecules in a similar way. With a suitable computer and data base, they will be able to identify proteins by reading their amino acid sequences.

Consider a complex and capable repair system. A volume of two cubic microns—about 2/1000 of the volume of a typical cell—will be enough to hold a central data base system able to:

1. Swiftly identify any of the hundred thousand or so different human proteins by examining a short amino acid sequence.
2. Identify all the other complex molecules normally found in cells.
3. Record the type and position of every large molecule in the cell.

Each of the smaller repair devices (of perhaps thousands in a cell) will include a less capable computer. Each of these computers will be able to perform over a thousand computational steps in the time that a typical enzyme takes to change a single molecular bond, so the speed of computation possible seems more than adequate. Because each computer will be in communication with a larger computer and the central data base, the available memory seems adequate. Cell repair machines will have both the molecular tools they need and "brains" enough to decide how to use them.

Such sophistication will be overkill (over cure?) for many health problems. Devices that merely recognize and destroy a specific kind of cell, for example, will be enough to cure a cancer. Placing a computer network in every cell may seem like slicing butter with a chain saw, but having a chain saw available does provide assurance that even hard butter can be sliced. It seems better to show too much than too little, if one aims to describe the limits of the possible in medicine.

Some Cures

The simplest medical applications of *nanomachines* will involve not repair but selective destruction. Cancers provide one example; infectious diseases provide another. The goal is simple: one need only recognize and destroy the dangerous replicators, whether they are bacteria, cancer cells, viruses, or worms. Similarly, abnormal growths and deposits on arterial walls cause much heart disease; machines that recognize, break down, and dispose of them will clear arteries for more normal blood flow. Selective destruction will also cure diseases such as herpes in which a virus splices its genes into the DNA of a host cell. A repair device will enter the cell, read its DNA, and remove the addition that spells "*herpes.*"

Repairing damaged, cross-linked molecules will also be fairly straightforward. Faced with a damaged, cross-linked protein, a cell repair machine will first identify it by examining short amino acid sequences, then look up its correct structure in a data basc. The machine will then compare the protein to this blueprint, one amino acid at a time. Like a proofreader finding misspellings and strange characters (characters), it will find any changed amino acids or improper cross-links. By correcting these flaws, it will leave a normal protein, ready to do the work of the cell.

Repair machines will also aid healing. After a heart attack, scar tissue replaces dead muscle. Repair machines will stimulate the heart to grow fresh muscle by resetting cellular control mechanisms. By removing scar tissue and guiding fresh growth, they will direct the healing of the heart.

This list could continue through problem after problem (*Heavy metal poisoning*? —Find and remove the metal atoms) but the conclusion is easy to summarize. Physical disorders stem from misarranged atoms; repair machines will be able to return them to working order, restoring the body to health, Rather than compiling an endless list of curable diseases (from *arthritis*, *bursitis*, cancer, and dengue to yellow fever and zinc chills and back again), it makes sense to look for the limits to what cell repair machines can do. Limits do exist.

Consider stroke, as one example of a problem that damages the brain. Prevention will be straightforward: Is a blood vessel in the brain weakening, bulging, and apt to burst? Then pull it back into shape and guide the growth of reinforcing fibers. Does abnormal cloning threaten to block circulation? Then dissolve the clots and normalize the blood and blood-vessel linings to prevent a recurrence. Moderate neural damage from stroke will also be repairable: if reduced circulation has impaired function but left cell structures intact, then restore circulation and repair the cells, using their structures as a guide in restoring the tissue to its previous state. This will not only restore each cell's function, but will preserve the memories and skills embodied in the neural patterns in that part of the brain.

Repair machines will be able to regenerate fresh brain tissue even where damage has obliterated these patterns. But the patient would lose old memories and skills to the extent that they resided in that pan of the brain. If unique neural patterns are truly obliterated, then cell repair machines could no more restore them than art conservators could restore a tapestry from stirred ash. Loss of information through obliteration of structure imposes the most important, fundamental limit to the repair of tissue.

Other tasks are beyond cell repair machines for different reasons—maintaining mental health, for instance. Cell repair machines will be able to correct some problems, of course. Deranged thinking some times has biochemical causes, as if the brain were drugging or poisoning itself, and other problems stem from tissue damage. But many problems have little to do with the health of nerve cells and everything to do with the health of the mind.

A mind and the tissue of its brain are like a novel and the paper of its book. Spilled ink or flood damage may harm the book, making the novel difficult to read. Book repair machines could nonetheless restore physical "*health*" by removing the foreign ink or by drying and repairing the damaged paper fibers. Such treatments would do nothing for the book's *content*, however, which in a real sense is nonphysical. If the book were a cheap romance with a moldy plot and empty characters, repairs would be needed not on the ink and paper, but on the novel. This would call not for physical repairs, but for more work by the author, perhaps with advice.

Similarly, removing poisons from the brain and repairing its nerve fibers will thin some mental fogs, but not revise the content of the mind. This can be changed by the patient, with effort; we are all authors of our minds, But because minds change themselves by changing their brains, having a healthy brain will aid sound thinking more than quality paper aids sound writing. Readers familiar with computers may prefer to think in terms of hardware and software. A machine could repair a computer's hard ware while neither understanding nor changing its software.

Such machines might stop the computer's activity but leave the patterns in memory intact and ready to work again. In computers with the right kind of memory (called *nonvolatile*), users do this by simply switching off the power. In the brain the job seems more complex, yet there could be medical advantages to inducing a similar state.

Anesthesia Plus

Physicians already stop and restart consciousness by interfering with the chemical activity that underlies the mind. Throughout active life, molecular machines in the brain process molecules. Some disassemble sugars, combine them with oxygen, and capture the energy this releases. Some pump salt ions across cell membranes; others build small molecules and release them to signal other cells. Such processes make up the brain's metabolism, the sum total of its chemical activity. Together with its electrical effects, this metabolic activity underlies the changing patterns of thought.

Surgeons cut people with knives. In the mid-1800s, they learned to use chemicals that interfere with brain metabolism, blocking conscious thought and preventing patients from objecting so vigorously to being cut. These chemicals are *anesthetics*. Their molecules freely enter and leave the brain, allowing anesthetists to interrupt and restart human consciousness.

People have long dreamed of discovering a drug that interferes with the metabolism of the entire body, a drug able to interrupt metabolism completely for hours, days, or years. The result would be a condition of *biostasis* (from *bio*, meaning life, and *stasis*, meaning a stoppage or a stable state). A method of producing *reversible biostasis* could help *astronauts* on long space voyages to save food and avoid boredom, or it could serve as a kind of one-way time travel. In medicine, *biostasis* would provide a deep *anesthesia* giving physicians more time to work. When emergencies occur far from medical help, a good biostasis procedure would provide a sort of universal first-aid treatment: it would stabilize a patient's condition and prevent molecular machines from running amok and damaging tissues.

But no one has found a drug able to stop the entire metabolism the way anesthetics stop consciousness—that is, in a way that can be reversed by simply washing the drug out of the patient's tissues. Nonetheless, *reversible biostasis* will be possible when repair machines become available.

To see how one approach would work, imagine that the blood stream carries simple molecular devices to tissues, where they enter the cells. There they block the molecular machinery of metabolism—in the brain and elsewhere—and tie structures together with stabilizing cross-links. Other molecular devices then move in, displacing water and packing themselves solidly around the molecules of the cell. These steps stop metabolism and preserve cell structures. Be cause cell repair machines will be used to reverse this process, it can cause moderate molecular damage and yet do no lasting harm. With metabolism stopped and cell structures held firmly in place, the patient will rest quietly, dreamless and unchanging, until repair machines restore active life.

If a patient in this condition were turned over to a present-day physician ignorant of the capabilities of cell repair machines, the consequences would likely be grim. Seeing no signs of life, the physician would likely conclude that the patient was dead, and then would make this judgment a reality by "*prescribing*" an *autopsy*, followed by burial or *burning*.

But our imaginary patient lives in an era when biostasis is known to be only an interruption of life, not an end to it. When the patient's contract says "*wake me*!" (or the repairs are complete, or the flight to the stars is finished), the attending physician begins resuscitation. Repair machines enter the patient's tissues, removing the packing from around the patient's molecules and replacing it with water. They then remove the cross-links, repair any damaged molecules and structures, and restore normal concentrations of salts, blood sugar, ATP, and so forth. Finally, they unblock the metabolic machinery. The interrupted metabolic processes resume, the patient yawns, stretches, sits up, thanks the doctor, checks the date, and walks out the door.

From Function to Structure

The reversibility of biostasis and irreversibility of severe stroke damage help to show how cell repair machines will change medicine. Today, physicians can only help tissues to heal themselves. Accordingly, they must try to preserve the *function* of tissue. If tissues cannot function, they cannot heal. Worse, unless they are preserved, deterioration follows, ultimately obliterating structure. It is as if a mechanic's tools were able to work only on a running engine.

Cell repair machines change the central requirement from preserving *function* to preserving *structure*. Repair machines will be able to restore brain function with memory and skills intact only if the distinctive structure of the neural fabric remains intact. Biostasis involves preserving neural structure while deliberately blocking function. All this is a direct consequence of the molecular nature of the repairs. Physicians using scalpels and drugs can no more repair cells than someone using only a pickax and a can of oil can repair a fine watch. In contrast, having repair machines and ordinary nutrients will be like having a watchmaker's tools and an unlimited supply of spare parts. Cell repair machines will change medicine at its foundations.

From Treating Disease to Establishing Health

Medical researchers now study diseases, often seeking ways to pre vent or reverse them by blocking a key step in the disease process.

The resulting knowledge has helped physicians greatly: they now prescribe insulin to compensate for diabetes, antihypertensive to prevent stroke, penicillin to cure infections, and so on down an impressive list. Molecular machines will aid the study of diseases, yet they will make understanding disease far less important. Repair machines will make it more important to understand health.

The body can be ill in more ways than it can be healthy. Healthy muscle tissue, for example, varies in relatively few ways: it can be stronger or weaker, faster or slower, have this antigen or that one, and so forth. Damaged muscle tissue can vary in all these ways, yet also suffer from any combination of strains, tears, viral infections, parasitic worms, bruises, punctures, poisons, sarcomas, wasting diseases, and congenital abnormalities. Similarly, though neurons are woven in as many patterns as there are human brains, individual synapses and dendrites come in a modest range of forms—if they are healthy.

Once biologists have described normal molecules, cells, and tissues, properly programmed repair machines will be able to cure even unknown diseases. Once researchers describe the range of structures that (for example) a healthy liver may have, repair machines exploring a malfunctioning liver need only look for differences and correct them. Machines ignorant of a new poison and its effects will still recognize it as foreign and remove it. Instead of fighting a million strange diseases, advanced repair machines will establish a state of health.

Developing and programming cell repair machines will require great effort, knowledge, and skill. Repair machines with broad capabilities seem easier to build than to program. Their programs must contain detailed knowledge of the hundreds of kinds of cells and the hundreds of thousands of kinds of molecules in the human body. They must be able to map damaged cellular structures and decide how to correct them. How long will such machines and programs take to be developed? Offhand, the state of biochemistry and its present rate of advance might suggest that the basic knowledge alone will take centuries to collect. But we must beware of the illusion that advances will arrive in isolation.

Repair machines will sweep in with a wave of other technologies. The assemblers that build them will first be used to build instruments for analyzing cell structures. Even a pessimist might agree that human biologists and engineers equipped with these tools could build and program advanced cell repair machines in a hundred years of steady work. A cocksure, far-seeing pessimist might say a thousand years. A really committed nay-sayer might declare that the job would take people a million years. Very well: fast technical AI systems—a million fold faster than scientists and engineers—will then develop advanced cell repair machines in a single calendar year.

A Disease Called "Aging"

Aging is natural, but so were smallpox and our efforts to prevent it. We have conquered smallpox, and it seems that we will conquer aging. *Longevity* has increased during the last century, but chiefly be cause better sanitation and drugs have reduced bacterial illness. The basic human lifespan has increased little.

Still, researchers have made progress toward understanding and slowing the aging process. They have identified some of its causes, such as uncontrolled cross-linking. They have devised partial treatments, such as *antioxidants* and *free-radical inhibitors*. They have proposed and studied other mechanisms of aging, such as "*clocks*" in the cell and changes in the body's hormone balance. In laboratory experiments, special drugs and diets have extended the lifespan of mice by 25 to 45 percent.

Such work will continue; as the baby boom generation ages, expect a boom in aging research. One biotechnology company, Senetek of Denmark, specializes in *aging research*. In April 1985, Eastman Kodak and ICN Pharmaceuticals were reported to have joined in a $45 million venture to produce *isoprinosine* and other drugs with the potential to extend lifespan. The results of conventional antiaging research may substantially lengthen human life spans and improve the health of the old—during the next ten to twenty years. How greatly will drugs, surgery, exercise, and diet extend lifespans? For now, estimates must remain guesswork. Only new scientific knowledge can rescue such predictions from the realm of speculation, because they rely on new science and not just new engineering.

With cell repair machines, however, the potential for life extension becomes clear. They will be able to repair cells so long as their distinctive structures remain intact, and will be able to replace cells that have been destroyed. Either way, they will restore health. Aging is fundamentally no different from any other physical disorder; it is no magical effect of calendar dates on a mysterious life-force. Brittle bones, wrinkled skin, low enzyme activities, slow wound healing, poor memory, and the rest all result from damaged molecular machinery, chemical imbalances, and misarranged structures. By restoring all the cells and tissues of the body to a youthful structure, repair machines will restore youthful health.

People who survive intact until the time of cell repair machines will have the opportunity to regain youthful health and to keep it almost as long as they please. Nothing can make a person (or any thing else) last forever, of course, but barring severe accidents, those wishing to do so will live for a long, long time. As a technology develops, there comes a time when its principles become clear, and with them many of its consequences. The principles of rocketry were clear in the 1930s, and with them the consequence of spaceflight. Filling in the details involved designing and testing tanks, engines, instruments, and so forth. By the early 1950s, many details were known. The ancient dream of flying to the Moon had became a goal one could plan for.

The principles of molecular machinery are already clear, and with them the consequence of cell repair machines. Filling in the details will involve designing molecular tools, assemblers, computers, and so forth, but many details of existing molecular machines are known today. The ancient dream of achieving health and long life has be come a goal one can plan for.

Medical research is leading us, step by step, along a path toward molecular machinery. The global competition to make better materials, electronics, and biochemical tools is pushing us in the same direction. Cell repair machines will take years to develop, but they lie straight ahead. They will bring many abilities, both for good and for ill. A moment's thought about military *replicators* with abilities like those of cell repair machines is enough to turn up nauseating possibilities.

10

Genomic Weapons

Biological weapons are not a recent invention and man has been using them since the Middle Ages. It is only since the 50s, however, that we have been using our advanced biological knowledge to manipulate viruses, bacteria and toxins in order to make them more harmful and turn them into proper weapons. Perhaps a new phase in the history of biological weapons is beginning today: the genomic phase. Our knowledge about DNA is being used to build new, increasingly powerful weapons: pathogens genetically modified to be even more deadly, to resist any treatment, and be nothing less than invincible.

Claire Fraser and Malcom Dando gave voice to the fear that technology of this kind could end up in the hands of terrorists-scientists in a commentary published in *Nature Genetics* just after 9/11. Their article recalled the theory of viruses that can remain sleeping in the victims' DNA and wake only in specific conditions, such as the release of special chemical substances in the air. It recalled the idea of terrorists using a new kind of anti-cancer molecules able to induce cells to self-disruption, and that of "*ethnic weapons*": germs that would be lethal only for populations with specific genetic characteristics.

From Labs to the Congress

To date, scenarios of this kind may still be considered science-fiction; but, in fact, someone could be willing to try it, sooner or later. What is more is that new biological weapons can even be created by chance. In early 2001, for example, an Australian research team involuntarily selected a viral strain of mousepox that could kill also vaccinated mice. Someone wondered: could the same thing be possible for the human strain. In August 2002, biologists Eckart Wimmer,

Jeronimo Cello and Aniko Paul, of New York State University, wrote on Science that you could synthesize poliovirus cDNA using information that you could easily find on the Internet, provided the necessary equipment and knowledge, of course.

However, after the publication of their article, US representatives had invited the scientific community to take on security measures to prevent terrorist groups or enemy States from gaining easy access to useful information and building mass weapons. On January 9, 2003, publishers, scientists, public security experts and Government representatives met at the National Academy of Science to find an agreement on how to combine security and free access to scientific information. A common decision was taken by the major scientific journals and institutes the following day and was announced, on February 15, in Denver, during the annual meeting of the American Association for the Advancement of Science: self-governance of all articles that could contain useful information for bioterrorists.

Preventive Self-governance

The publishers who signed the Statement marked that the decision of censoring publications also meant protecting their own publishing freedom. The main fear, mostly on the western side of the Atlantic Ocean, is that a stricter censoring would be exercised over scientific journals if they did not opt for self-governance. "We're aware that if we do not police ourselves then Congress may do it for us" Eckart Wimmer, the poliovirus synthesis biologist and co-signer of the Denver Statement, told Nature. This fear is not as far-fetched as it may seem, as, already in February 2002, the US Department of Defence had announced their intention to review all biomedical papers before publication. They never really reached this point, but, by the end of 2002, many scientific documents had vanished from US public data banks. Among these documents were also a few old vintage studies dating back to the 40s which some reforms had made free to public access in order to give greater transparency to the Government action.

Actually, some self-censoring episode had already taken place in the US. Ron Atlas, president of the American Society for Microbiology, one of the main supporters of the Denver Statement, said that, between 2001 and 2002, the 11 papers controlled by his society had accepted 134 reports on lethal biological agents, and that 2 of them had worried the editors and had been published only after some changes had been agreed upon with the authors. One of them had had the introduction removed, because it excessively stressed the harmfulness of a biological agent; the other had had some passages censored, as they explained how to modify a natural toxin so as to kill one million instead of "only" ten thousand people. "Scientists, publishers and directors of scientific magazines have the moral duty to work for the well-being of humanity. Taking every possible precaution to prevent the information produced and published by the scientific community from being used in improper ways is part of that moral duty" said Ron Atlas.

End of Innocence

But when can a publisher come to the conclusion that an article is potentially more dangerous than useful? Deciding a priori which pieces of information are harmful is no banal issue. Even the Denver Statement admits that, at the moment, no definition or list of delicate information that should not be published is available. "Seeing the border line between a research that will lead to peaceful applications of a scientific discovery and a research that will give you the recipe for a bomb is not easy" says also Joseph Rotblat, the nuclear physicist who abandoned the Manhattan Project of the first atomic bomb and who was awarded the Nobel Peace Prize "for his efforts to diminish the part played by nuclear arms in international politics" in 1995. "This is why many scientists of the most pertinent fields should be involved in this project. If we are careful about this, science communication will not suffer heavy consequences" he adds.

Who the censors will be is still uncertain, as are the fields involved; however, it is easy to foresee that the main target of censoring will be the methodological details. "These are the details every researcher reads first, because they indicate whether the experiment was carried out properly and how it is possible to replicate it", says Marcello Buiatti, a geneticist at the University of Florence. Some fear that undermining free access to this information will end up jeopardizing the very basis of science. "The fact is that no experiment can be considered scientific if not replicable. And no one can replicate an experiment without knowing the details of the original", Mark Frankel, responsible for the "Program on Scientific Freedom, Responsibility and Law" of the American Association for the Advancement of Science, told the English journal The Lancet, which did not sign the Denver Statement. Many people today, whether members of the scientific community or not, feel that without public communication of science there is no science at all. On this issue, however, Ron Atlas is quite reassuring: "The Statement we signed in Denver stresses the importance of the integrity of science. It explicitly forbids any censorship undermining scientific bases and states science must be reproducible by definition: details that are useful in this sense will not be censored. However, it is important that ethical considerations also become part of the peer review process, which has so far focused on the articles' quality and originality. Both the public and the financial supporters of scientific research should agree on this, thus rendering the consequences on scientific communications negligible".

Science Police

But who will be responsible of censoring articles? Will editors also filter information, as Ron Atlas hopes? "These people will, however, be chosen on an international level" Marcello Buiatti objects. "Will there be national, religious or ideological discriminations? And if they are not editors, who will they be? Will they be external experts, or perhaps part of this or that State's antiterrorism service?". These fears, as well, are not completely groundless: last April, at a London meeting on bioterrorism, John Steinbruner, an arms control expert at the University of Maryland, called for an "international body of scientists and public representatives who would authorize scientific research carrying potential for grave social consequences". His proposal has already been nicknamed "*science police*".

Obviously, however, the very ambiguity of judging the harmfulness of an article could mean that a very high price will be paid to self-censoring. The biggest risk is that researchers may fear not being able to publish their results, thus slowing down their own career, and could abandon some research fields—such as that of pathogenic microorganisms—leaving them to become the monopoly of military laboratories, just where a stronger public control would be needed.

Opposition: Gagged Scientists

The scientific community is not unanimous on this issue: the critiques focus on the effectiveness of this measure. Already in May 2002 the American geneticists Timothy Read and Julian Parkhill had underlined, in a letter to the journal Nature, that having access to genetic information is of absolutely no help to terrorists, because the real difficulties of creating a biological weapon are the techniques needed to select, grow and spread pathogenic germs. On the other hand, if scientists are prevented from consulting genetic data banks, preventive actions such as the creation of new drugs, new vaccines and surveillance systems will be undermined.

The problem of free access to information is not new to modern life sciences: scientists have been coping with difficulties deriving from the patenting of economically significant scientific discoveries for some years now. Nor is it the first time that the circulation of the results of scientific research is controlled by someone else. Just think of what happened to nuclear physics during the 40s, with the militarization of the Manhattan Project, and the entire period of the Cold War between the United States and the Soviet Union. The situation of biomedical sciences is different, however, because research in this field always has important direct or indirect effects on our health. The Public Library of Science, a San Francisco non-profit organization supporting free access to scientific publications, gave a very harsh judging of the Denver Statement and charged its signers with lack of far-sightedness: "The benefits and dangers that a new discovery holds are rarely immediately evident, and the discoveries that have brought the greatest benefits to society have often had the most overt potential for danger. The identification of the agent that causes a deadly disease, and the development of methods for culturing the deadly agent, can be viewed as critical steps toward development of a vaccine, or a cookbook for bioterrorists. It is naive to imagine that the censorship of scientific ideas and discoveries based on their foreseeable potential for destructive use would significantly diminish the danger of terrorism. Instead, limits on intellectual freedom and the free flow of scientific information would stifle the scientific creativity that is vital to our defense against terrorism and other, greater threats to human welfare".

Two Cultures

Donald Kennedy, director of Science and signer of the Denver Statement, in the leading article of February 21, speaks of two colliding cultures. He does not refer to the long-lasting, now obsolete conflict between the scientific culture and the humanities. Kennedy maintains that what we are witnessing today is a new conflict between the "*culture of free science*" and the "*culture of security*", between those who think that the fears about scientific research are irrational and derive from poor understanding of the issue, and those who think that scientists are not aware of the potential dangers deriving from their profession. It is, in fact, a conflict between those who maintain that scientific activities have to be controlled by society and those who support the independence of science from any external constraint.

The scientific community has been discussing fears deriving from potential risks of biotechnologies since 1975, when at the end of a famous Asilomar, California meeting, researchers opted for a moratorium, that is, a suspension of every activity, until security measures had been satisfied. However, the risks of biotechnologies became a politically and socially alarming issue only at the end of the 90s, when the first GMOs entered the market and the spectre of human cloning gave rise to increasing fears. Today the threat of bioterrorism wakes an old fear: that powerful technology can end up into the wrong hands. Brian Spratt, of the London Imperial College, as much as hopes that the perils of biological weapons will arouse an appropriate response from the academic world: an answer based on a shared scientific code of conduct that should be part of every student's training.

Roots of Fear

Biological weapons have always aroused a very strong sense of disgust. In ancient times people called them poisons, because they did not know that infectious diseases were caused by microorganisms. Poisons were banned from the war code by Greeks and Romans, they were banned by Saracens, because they were incompatible with the teachings of the Koran, and by the laws of the Indian Code of Manu, as they were considered inhuman weapons. Biological weapons continue to arouse fears today. It hardly matters that military strategists assure the public that terrorists will keep preferring conventional weapons like explosives, which are more effective and easy to use; or that statistics show that viruses and bacteria rarely cause any victim when actually used as terrorist. The Washington Henry Stimson Center, a non-profit organization working for peace and international security, has collected data showing that during the past 25 years (1975-2000) terrorist attacks with biological weapons have caused only 2 victims. Five more people were killed in Autumn 2001 in the US by mail-spread anthrax. This means that, during the past 28 years, bioterrorism has caused 7 deaths all over the world: an extraordinary low number of victims.

Then why are we so afraid? What led the major international scientific journals to the undertaking of such a clamorous measure as self-governance? The fact is that the very idea of terrorists using this terrible kind of weapon awakes fears that are very deeply rooted in the collective unconscious. Biological weapons remind us of biblical plagues, and terrorists with no face or name strongly recall the plague spreaders. As if this were not enough, after what happened on 9/11, many people now think anything possible. The terrorist attack against the Twin Towers and the anthrax mails have led to a completely new perception of the risk of mass weapons. People now think that rudimental equipment and some recipes taken from the Internet are enough for anyone to build a biological weapon in his garage. The media, and not just the popular ones, have also contributed to the diffusion of this idea. Ian Roberts, pubic health expert at the London School of Hygiene and Tropical Medicine, states that the American medical journals have given too much room to bioterrorism, leading people to think a biological attack to be imminent, thus justifying the political necessity of a war in Iraq. "Actually, changing a virus or a bacterium into a biological weapon is a very complicated process: it requires abilities that no biology student has, and very expensive equipment that pharmaceutical industries, not university labs, dispose of" says Arturo Falaschi, director of the International Centre of Genetic Engineering and Biotechnologies (ICGEB) of Trieste, Italy. And he adds: "The success of a biological attack is a very remote chance, and editors know that. But in the collective psychosis of this period they have given up to those maintaining that sooner or later someone will try it out."

We ought not to forget that fear is also business, that fear can also produce money. Nature states that biotech companies, struggling for survival after 3 years of low investments, are now queuing outside the US Department of Treasury to get their share of the generous investments the Government is making on biosecurity. The year 2003 should see the investment of 6 billion dollars in long-term projects such as the development of new anthrax and smallpox vaccines, "*indispensable*" to face pandemics we will probably never see. "This is an opportunity created by fear" said Charles Cantor, an expert of bio-defence and scientific representative of Sequenom, a genetic company of San Francisco, California.

Conclusions

The bioterrorist threat and "*the culture of fear*" prevailing in western society after September 11 has led the scientific community to revise the ethical norms controlling scientific production and communication methods, and firstly the process of peer review. Self-governance is a sign of the deep change that is increasingly forcing scientists to come to terms with the needs, the wishes and the fears of the society they live in. This change, as it usually happens, is also affecting the communication processes by which a community, in this case the scientific one, defines its language and identity. This time it is not an internal process: it is society making explicit requests that call for a change in the usual ways of producing and communicating science, thus changing the idea society, itself, has of science.

11

Biological Weapons Capabilities

Five extensive databases were published in the 1990s covering nearly the entire 20th century, and several of these have been updated so as to remain current. It is extremely important to distinguish between the seven different categories of BW-related events that they cumulatively cover: hoaxes, threats, consideration or discussion of use, product tampering, purchase of materials, attacks on facilities, attempts to produce biological agents or attempts to use them, and actual use.

These databases were compiled by:

- Harvey McGeorge, 1994, covering 1945-94;
- Ron Purver, 1995, covering 1945-95;
- Bruce Hoffman, 1998, covering 1990-98;
- Seth Carus, 1999, covering 1990-99, and since updated; and,
- The Monterey Institute, 1999, covering 1990-99, and since updated.

The conclusions from these independent studies were uniform and mutually reinforcing. There is an *extremely low* incidence of real biological (or chemical) events, in contrast to the number of hoaxes, the latter spawned by administration and media hype since 1996 concerning the prospective likelihood and dangers of such events. A massive second wave of hoaxes followed the anthrax incidents in the United States in October-November 2001, running into global totals of tens of thousands. It is also extremely important that analysts producing tables of "*biological*" events not count hoaxes. A hoax is not a "*biological*" event, nor is the word "*anthrax*" written on a slip of paper the same thing as anthrax, or a pathogen, or a "*demonstration of threat*"—all of which various analysts and even government advisory groups have counted hoaxes as being on one occasion or another.

Those events that were real, and were actual examples of use, were overwhelmingly chemical, and even in that category, involved the use of easily available, off-the-shelf, nonsynthesized industrial products. Many of these were instances of personal murder, and not attempts at mass casualty use. The Sands/Monterey compilation indicated that exactly one person was killed in the United States in the 100 years between 1900 and 2000 as a result of an act of *biological* or *chemical terrorism*. Excluding the preparation of *ricin*, a plant toxin that is relatively easier to prepare, there are only a few recorded instances in the years 1900 to 2000 of the preparation or attempted preparation of pathogens in a private laboratory by a nonstate actor. The significant events to date are:

1. 1984, the Rajneesh, The Dalles, Oregon, use of *Salmonella* on food;
2. 1990-94, the Japanese Aum Shinrikyo group's unsuccessful attempts to procure, produce and disperse *anthrax* and *botulinum toxin*;
3. 1999, November 2001, al-Qaida, the unsuccessful early efforts to obtain *anthrax* and to prepare a facility in which to do microbiological work;
4. October-November 2001, the successful "*Amerithrax*" distribution of a high-quality dry-powder preparation of *anthrax spores*, which had been prepared within the preceding 24 months.

Before discussing the *Amerithrax* and *al-Qaida* experiences in some further detail, two books, one published and one still in press, should be mentioned. These are particularly important because they are collections of case studies which, between the two, contain detailed reviews of virtually all the groups or individuals who have—or who had been alleged to have—prepared or used chemical or biological agents. The first of the books is *Toxic Terror*, edited by Jonathan Tucker, and the second is the forthcoming *Motives, Means and Mayhem: Terrorist Acquisition and Use of Unconventional Weapons*, edited by John Parachini. Between them the two books report on 28 case studies. They demonstrate that several right-wing groups in the United States produced ricin by extraction from mashed castor bean pulp, and that the Rajneesh group did culture the *Salmonella* that it obtained. However, there is apparently no other "*terrorist*" group that is known to have successfully cultured any pathogen. It is precisely because of the exceptional nature of the Amerithrax case that it becomes crucially significant to identify the person or persons who made the *anthrax* preparation, and to determine whether that is the same individual or individuals who prepared and mailed the postal envelopes. We will return to this in a moment. In advance of the publication of his book, *Parachini* summarized the conclusions from those studies that "provide an empirical foundation to assess the motivations, behaviour, and patterns related to terrorist interest, or alleged interest, in *unconventional weapons*."

Perhaps the most important discovery from the first of the two books was that "Upon rigorous inspection, several of the empirical cases frequently cited in the media and scholarly literature proved to be apocryphal." Parachini then discusses several factors that appear to be most significant in understanding the case studies. He finds the mindset of the group leaders of the organization, exogenous and internal constraints, and a combination of opportunity and the technical capacity of the group to be "...the factors that most significantly influence a group's propensity to seek to acquire and to use *unconventional weapons*."

These conclusions are consistent with those made by another highly experienced terrorism specialist, Dr. Yoram Schweitzer of the Jaffee Center for Strategic Studies, Tel Aviv University. In a recent conference presentation, he enumerated four factors which he felt served as inhibitions to the consideration of biological weapons within terrorist organizations: state dependency, requirements of their own local constituency, requirements of the international constituency, and group survival. To the degree that the leadership of a particular terrorist organization does not have or escapes from these considerations, they may consider the use of *biological agents* more seriously. This may explain why the experience of the Rajneesh, *Aum Shinrikyo*, and *al-Qaida* groups followed a path different from other terrorist groups. Similar conceptions were explored as far back as 1989 in a RAND study authored by Dr. Jeffrey Simon.

This sort of analysis of real cases of the relevant behaviour of real terrorist groups, carried out by experienced analysts of terrorism, when it is not disregarded entirely, is frequently met with disdain by the proselytizers of "the *bioterrorist threat*," as an example presented later in this monograph indicates. The extremely brief entry regarding terrorist groups and BW and CW capabilities that appeared in the December 2004 report of the U.S. National Intelligence Council was again minimalist in its description: Given the goal of some terrorist groups to use weapons that can be employed surreptitiously and generate dramatic impact, we expect to see terrorist use of some readily available biological and chemical weapons.

Since this is a report that was looking ahead to the period between 2005 and 2020—the next 15 years—the reference to "*readily available*" materials is particularly notable. It harks back to the conclusions of the 20th century database studies, and certainly does not seem to anticipate efforts at synthesis, genetic engineering, or anything beyond the most elementary products. Dr. Stephen Morse predicted much the same speaking to a day-long conference convened by the U.S. National Academy of Sciences in January 2003. Morse previously served in the Defense Advanced Research Projects Agency (DARPA) of the U.S. Department of Defense (DoD), and currently is director of the Center for Public Health Preparedness at Columbia University's School of Public Health. In answering his own question "What sources would terrorists use," he stated that "Most are likely to use easily obtained materials," and that "state-sponsored [terrorists] might use '*classical BW*'."

The latest U.S. Government intelligence estimates of the BW capabilities of terrorist/nonstate actor groups became available in February and March 2005. They first appeared in three presentations to the U.S. Senate Select Committee on Intelligence on February 16, 2005. Porter Goss, Director of Central Intelligence: "It may be only a matter of time before *al-Qaida* or another group attempts to use *chemical*, *biological*, *radiological*, and *nuclear weapons* (CBRN)."

Robert Mueller, Director of the Federal Bureau of Investigation: "...we are concerned that they are seeking weapons of mass destruction including chemical weapons, so-called '*dirty bombs*' or some type of biological agent such as *anthrax*.... I am also very concerned with the growing body of sensitive reporting that continues to show *al-Qaida's* clear intention to obtain and ultimately use some form of *chemical*, *biological*, *nuclear* or *high-energy explosives* (CBRNE) material in its attacks against America."

Jim Loy, Deputy Secretary, U.S. Department of Homeland Security: "...the most severe threats revolve around al-Qaida and its affiliates' long-standing intention to develop, procure, or acquire chemical, biological, radiological, and even nuclear, weapons for mass-casualty attacks. Al-Qaida and affiliated elements currently have the capability to produce small amounts of crude biological toxins and toxic chemical materials, and may have acquired small amounts of radioactive materials."

On March 17, 2005, it was the turn of Vice Admiral Jacoby, Director of the U.S. Defense Intelligence Agency, in a presentation to the U.S. Senate Armed Services Committee: We judge terrorist groups, particularly *al-Qaida*, [to] remain interested in *Chemical, Biological, Radiological* and *Nuclear* (CBRN) weapons. Al-Qaida's stated intention to conduct an attack exceeding the destruction of 9/11 raises the possibility that planned attacks may involve unconventional weapons. There is little doubt it has contemplated using radiological or nuclear material. The question is whether al-Qaida has the capability. Because they are easier to employ, we believe terrorists are more likely to use biological agents such as ricin or botulinum toxin or toxic industrial chemicals to cause casualties and attack the psyche of the targeted populations.

CIA Director Porter Goss made a presentation to the Committee as well, and repeated the exact wording of his February 16 remarks quoted above. These rather similar extracts are the only references to nonstate actor interest or capability in the CBW area in all the presentations. They are significant for several reasons:

1. No other group than *al-Qaida* was mentioned.
2. All of the statements are less specific, more general, lower key, than in the preceding (Tenet) years.
3. In two of the three statements, B is grouped together with C, R and N, and in one even with "E" (high-energy explosives), making it impossible for someone without prior knowledge to know whether there is a specific basis for including B.
4. One of the statements refers to "small amounts of crude *biological toxins*" which undoubtedly refers to ricin and not to any pathogen.
5. It has been known at least since November 2001, when U.S. and UK military forces occupied Afghanistan, that al-Qaida had been "seeking...*anthrax*" for 2 or 3 years prior to that date. The information therefore concerns al-Qaida activities before the fall of Afghanistan, and there is no publicly available information to indicate that the group has been able to continue their efforts since the end of 2001.

Anthrax is endemic in Afghanistan, and a limited domestic animal vaccination program existed in the country. Nevertheless, there were undoubtedly some number of animal cases of the disease per year, but the group apparently never obtained a *pathogenic strain*.

In addition, all of the four recent specific references to al-Qaida-affiliated groups and ricin have turned out to be spurious. When the Al Ansar camp in Northeast Iraq, "*Kermal*," was overrun by U.S. military forces, sampling showed no presence whatsoever of ricin, nor of materials for its preparation. (In fact, the camp lacked running water.) Before the invasion, there had been several statements by U.S. Government and military officials stating that the group was preparing ricin in the camp. As late as the Vice Presidential debate on October 5, 2004, U.S. Vice President Cheney, referring to Abu Musab al-Zarqawi, said, "He set up shop in Baghdad, where he oversaw the poisons facility up at Kermal, where the terrorists were developing ricin and other deadly substances to use." Since this remark was made long after the U.S. Government knew that the camp had contained no ricin, Cheney's statement has to be considered either ignorance or fabrication. There is no mention whatsoever of the Al Ansar camp in the report of the Iraq Survey Group.

Individuals arrested by French police did not have any ricin nor had they prepared any; they had only "*planned to*" produce ricin. The "*chemicals*" that the group had was apparently sodium cyanide. The "*plots*" in the spring of 2003 by a group associated with Abu Musab Zarqawi in Jordan reportedly again involved sodium cyanide, and not ricin. (Since Jordanian authorities have referred to the planned use of 20 tons of high explosives in the planned event, there is every reason to assume that the *cyanide* would have been destroyed in any such massive high explosives blast.)

The London case is the last, and it is interesting to look at it in some detail. The group, arrested in Wood Green, London, was in possession of 22 *castor bean seeds*. Their equipment was a coffee grinder, "...with a brown residue" (probably of coffee), a mortar and pestle, and a hand-written recipe taken off the internet at an internet cafe and transcribed into Arabic. The recipe was a derivative of the Maxwell Hutchkinson recipe in the notorious *The Poisoner's Handbook* sold in thousands of copies at U.S. gun shows, a recipe that would very likely not produce *ricin* or extremely little of it.

The first tests for *ricin* in the London apartment were done by a field test kit and apparently registered positive. Within 2 days, 20 more specific tests were carried out at the British Defense Science Technology Laboratory, Porton Down, resulting in 17 "*negatives*" and three false positives. However the task of informing the London Metropolitan Police fell to another Porton staffer with public liaison responsibilities, who apparently either did not understand or confused the information that he was to relay, with the result that he phoned the press and police, saying that "*traces of ricin*" had been found. His actions were later attributed to "*incompetence.*" Despite this, as late as mid-February 2005, an official UN investigative group reported to the UN Security Council that "al-Qaida-associated groups in both the United Kingdom and Jordan came close to mounting such [C, B, R, or N] attacks. It seems only a matter of time before a successful *chemical*, *biological*, *radiological* and *nuclear attack* occurs." (All four at once!) All of the above is incorrect. No *ricin* was found. Whether the UK Metropolitan Police ever announced a public correction in 2003 is not known. It is unfortunate that misinformation of this sort found its way into a UN document.

Scientists at Porton Down subsequently followed the Hutchkinson recipe as an experiment. It produced sufficient *ricin* to kill one person if the total quantity would have been *injected* into a person. If eaten, it was sufficient to have caused *vomiting* and *abdominal pain*. If applied to doorknobs to act as a contact poison, it would have had no effect at all. Three other "*recipes*" were found with this ricin "*recipe*": a "*rotten meat poison*" to be prepared by mixing "*corn flour*, *meat*, *dung*, and *dust* together in a can," a "*recipe*" for "*cyanide poison*" to be obtained by boiling many thousands of ground apple seeds, and a "*potato poison*" of equal inutility. However, Peter Clarke, the head of Britain's antiterrorist police branch, described the poison plot as being "*highly serious*," while Nigel Sweeney, the prosecutor, said "These were no playtime recipes. These are recipes that experts give credence to and experiments show work. They are scientifically viable and potentially deadly."

Both statements are clearly wrong, and there is no reason that law enforcement officials anywhere in the world should be able to do their work without inaccurate and sensationalist comments. As for the "*Encyclopedia of Jihad*," in which such rudimentary and often inadequate recipes are supposedly located, it is not clear when it was composed. One suggestion is that its origin goes back to the 1978-88 period of Afghan resistance to the USSR, which would be as much as 10 years before al-Qaida came into existence. A second suggestion is that it is an agglomeration that grew over the years, with material successively added to it, materials on toxin production being added in later years. Although the "*Encyclopedia*" is routinely attributed to *al-Qaida*, it therefore may rather have been inherited, or adopted, by the group.

Al-Qaida BW Efforts in Afghanistan: 1997-98 to 2001

When we move to the *al-Qaida* group in Afghanistan, the picture rapidly becomes much more serious, and all the preceding semi-farcical events can be seen as inconsequential trivia. The first significant and meaningful information on what *al-Qaida* may have hoped at some point to achieve in the area of BW appeared in a single page in the journal *Science* in mid-December 2003, and then in declassified documentary materials that were obtained in the last week of March 2004. Appended to the single page in *Science* by a computer link was a list of 32 items: 11 books and 21 professional journal papers nearly all dating from the 1950s and 1960s dealing with pathogens or with BW. These were found in an *al-Qaida* training camp near Kandahar, Afghanistan, in December 2001. Half of the books dealt with historical or general aspects of BW and would be of little operational utility to an effort to produce BW agents. However, at least some of the journal papers and the remaining half of the books could be useful for such an effort. A note in the *Science* paper identified these as the documents referred to by CIA Director Tenet in his February 2002 Senate testimony quoted below. They were found only a few kilometers from the site near Kandahar airport which contained the rudimentary equipment also procured by al-Qaida.

Most important of all, the documents indicated that "...*al-Qaida's* BW initiative included recruitment of individuals with Ph.D.-level expertise who supported planning and acquisition efforts by their familiarity with the scientific community." The journal papers concerned B. anthracis and Clostridium botulinum, but also Yersinia pestis (*plague*) and *Hepatitis* A and C. Fragments of two of the classified materials were included in the *Science* article as photocopies of handwritten letters. The letterhead of one of them read "Society for Applied Microbiology." The second item reports that the individual was not able to obtain a pathogenic culture of anthrax, and that "the culture available in [deleted] is nonpathogenic." The webpage of the Society for Applied Microbiology advertises that organization as "the UK's oldest microbiological society." Another snippet from the handwritten letter, which explained that its author would "require at least the air ticket expenses," indicated that the person was flying either to or from the UK. The letter fragment also explained that "The money with me is only to buy strains of *vaccines*."

When the classified documents were obtained, it turned out that nearly all of the pages consisted of the journal articles themselves, as well as medical handbook excerpts on *anthrax*, *plague*, *botulinum*, etc. There were also many additional pages of references to both books and journals, including many standard reference works such as the SIPRI volumes on *The Problems of Chemical and Biological Warfare* and the 1969 UN study on chemical and biological weapons. It was the remaining 10 pages that were of importance. They were two 3-page letters and accompanying handwritten notes suggesting the layout of a laboratory and the equipment recommended to outfit it, and "*program requirements*" including the time needed to train whoever was going to work in the laboratory, and that person's assistants.

The correspondent already had a Ph.D. It is clear that the author of the letter was not a native English speaker, and annotations on some of the papers that he sent from England were in Arabic. Yet the letters and the accompanying notes were written in English. It suggested that either English was the only language that he and the recipient shared in common, or it was the preferred language for them to use with each other. In fact, the author was a Pakistani microbiologist, whose native language would be Dari, and there is reason to believe that he was writing to the Egyptian, Dr. al-Zawahiri, Osama bin Laden's deputy. The writer reports visiting a BL-3 facility—apparently in the UK—at which time he had been shown a pathogen collection. He was not only trying to obtain and export pathogen cultures, but he was also seeking to buy vaccines for protecting personnel against anthrax infection. He was being supplied by *al-Qaida* with funds with which to buy equipment and materials, which he itemized. He had also been attending various European conferences dealing with pathogens—or had obtained their proceedings—including a conference on *anthrax*. The latest dates of these were in July and September 1999. He had signed his letters, named the laboratory that he had visited, and named its laboratory director and identified one or two other individuals by name. All these identifications were deleted in the declassified materials.

What the documents indicated was an individual with Ph.D.-level training, who understood the professional microbiology literature, and who understood professional procedures for purchasing pathogen cultures. He was willing to trade on the access provided by his status, while concealing the true purpose of his activities, which was to provide *al-Qaida* with the means to attempt its first real BW production capability. However, he was not prepared to do any of the laboratory work himself. There is no evidence in any of the declassified pages to indicate that any bacterial cultures had yet been obtained, or that any had been shipped to Afghanistan or Pakistan, or that any work had yet begun. In fact, all the phrasing on these pages suggests that none of these things had yet occurred. There is also no mention of the procurement of bacterial culture media that would be necessary to have in hand before any work could begin.

These materials were characterized by various senior U.S. officials in 2002 and 2003. On February 25, 2002, General Tommy R. Franks, the commander of U.S. military forces in Afghanistan, reported that following the examination of over 110 sites in Afghanistan: ...the United States has yet to find evidence that *al-Qaida* was able to create a chemical or biological weapon at any of its camps, command centers, or caves in Afghanistan... We have seen evidence that al-Qaida had a desire to weaponize chemical and biological capability, but we have not yet found evidence that indicates that they were able to do so.

Similarly in February 2002, U.S. CIA Director Tenet stated...we know that al-Qaida was working to acquire some of the most dangerous chemical agents and toxins. Documents recovered from al- Qaida facilities in Afghanistan show that Bin Laden was pursuing *a sophisticated biological weapons research program.*

In his analogous "*Threats*" assessment in 2003, Tenet told the U.S. Senate Committee on Armed Services, "I told you last year... that bin Laden has a sophisticated biological weapons capability.... In Afghanistan, al-Qaida succeeded in acquiring both the expertise and equipment needed to grow biological agents, including a dedicated laboratory in an isolated compound in Kandahar."

The inclusion of the words "*pursuing*" and "*research program*" in 2002 arguably makes the statement factually correct. "*Pursuing*" is not the same as saying that *al-Qaida* had produced any usable product. However, Tenet's 2003 language is substantially different and implies much more, stating that *al-Qaida* "has a sophisticated biological weapons *capability*" [emphasis added]. Nevertheless, the declassified materials permitted some proper understanding for the first time of the basis for statements by U.S. Government officials that the status of *al-Qaida's* BW program might be more advanced than had been anticipated.

The relevant passages in the annual CIA and Defense Intelligence Agency (DIA) threat assessment presentations to the U.S. Senate in February 2004 also very likely reflected judgments based on the materials described above. From the Director of the DIA: *Al-Qaida* and other terrorist groups remain interested in acquiring *Chemical*, *Biological*, *Radiological*, and *Nuclear* (CBRN) *weapons*. We remain concerned about rogue scientists and the potential that state actors arc providing, or will provide, technological assistance to terrorist organizations. . . . While we have no intelligence suggesting states are planning to give terrorist groups these weapons, we remain concerned about, and alert to, the possibility.

And from the Director of the CIA: ...I have consistently warned this committee of al-Qaida's interest in *chemical*, *biological*, *radiological*, and *nuclear weapons*. Acquiring these remains a "religious obligation" in Bin Ladin's eyes, and al-Qaida and more than two dozen other terrorist groups are pursuing CBRN materials.... Although gaps in our understanding remain, we see al-Qaida's program to produce anthrax as one of the most immediate terrorist CBRN threats we are likely face. The report of the U.S. September 11, 2001 (9/11) Commission includes a bare few lines on the single individual identified to date, a Malaysian, who was to have carried out al-Qaida's laboratory work: In 2001, [Yazid] Sufaat would spend several months attempting to cultivate anthrax for al-Qaida in a laboratory he helped set up near the Kandahar airport.... Sufaat did not start on the al-Qaida biological weapons program until after the JI's [Jemaah Islamiah] December 2000 Church bombings in Indonesia, in which he was involved.

Yazid Sufaat was arrested in Malaysia in December 2001. Publicly available information about him has come from two important al- Qaida sources. The first was Khaled Sheikh Mohammed who was arrested on March 1, 2003, in Rawalpindi, Pakistan, at the home of a fugitive Pakistani bacteriologist, Dr. Abdul-Quddis Khan. Handwritten notes and computer hard drives were seized in the home, showing, according to a reporter's description, that al-Qaida had "...completed plans and obtained the materials required to manufacture two biological toxins—*botulinum* and *salmonella*—and the *chemical poison cyanide*." Cyanide would not be "*manufactured*," and it is ambiguous if the "materials required" were the pathogen cultures, the bacterial growth media, equipment needed, or which of the above. "*Plans*" does not indicate that any production took place. The press report of these discoveries was contradictory in places, but claimed that the recruitment of named scientists was discussed in the materials seized, production steps were outlined, and equipment, such as that found in Afghanistan, was described. Among items found was "a direction to purchase" *Bacillus anthracis*. Nothing so far translated indicated access to the most dangerous microbial strains or to any advanced processing or delivery methods.

Mohammed also told his interrogators that Sufaat "...took the lead in developing biological weapons for *al-Qaida* until he was arrested by Malaysian authorities." Sufaat reportedly obtained a Bachelors degree "in biological sciences," with a "clinical laboratory concentration" from California State University in Sacramento in 1987. He then served as a laboratory technician in the Malaysian military and in 1993 established a company in Malaysia "to test the blood and urine of foreign workers and state employees for drug use." In the course of recent years, his company, and possibly another owned by his wife, appear to have been involved in financial transfers and the purchase of *ammonium nitrate* for producing explosives on behalf of groups affiliated with *al-Qaida* operating in Indonesia, Malaysia, and the Philippines.

The suggestion that Sufaat was not able to procure an appropriate strain of anthrax for use as a pathogen demonstrates the same difficulty faced by the Aum Shinrikyo group in Japan, which was only able to obtain the veterinary vaccine strain of anthrax. This appears to have been corroborated by reports in October 2003. A photograph taken at an internationally supported animal vaccine production facility outside of Kabul by an Associated Press photographer in November 2001 showed a large glass carboy jar labeled "*Anthrax spore Concentration.*" It almost certainly contained Sterne strain *anthrax vaccine*. While incriminating as to al-Qaida's eventual intentions, all information to date indicates that al-Qaida could not possibly have been responsible for the anthrax attacks in the United States in 2001.

Additional reporting about Yazid Sufaat came from Hambali (Raduan Isamuddin), the Indonesian operative of the al-Qaida affiliated organization, Jemaah Islamiah, who was responsible for the *Bali bombing* attack in August 2003. After his capture, Hambali told his interrogators that he had earlier been collaborating with Sufaat, that he had been "trying to open an al-Qaida bio-weapons branch plant," and that Sufaat had been "working on an al-Qaida *anthrax program* in Kandahar," in Afghanistan, but that after the U.S. attack on the Taliban, they had planned to move the "program" to Indonesia. However, Sufaat had been unable to obtain a pathogenic strain of *anthrax*. In another report, U.S. and Malaysian security officials more accurately described the al-Qaida program to develop *biological* and *chemical weapons* as having been "in the early 'conceptual stage' when it was cut short by the U.S. invasion of Afghanistan."

CBS nevertheless reported this as "al-Qaida may be hard at work trying to produce weaponized anthrax and other biological weapons."

Two weeks later, rumors that Jemmah Islamiah branches in the Philippines were producing biological or chemical agents were quickly proved to be spurious. An important question is how much and what kind of actual laboratory work Sufaat might have been able to achieve in the "*several months*" available to him at the Kandahar site. Sufaat and Hambali apparently made four trips between Kandahar and Karachi to purchase materials. The 1,000+ kilometer distance each way means that these trips would have been by air, but nevertheless could have together required several weeks to a month. Other lower-ranking al-Qaida members also made purchasing trips to Pakistan. As best is known, the Kandahar site appears not to have yet been functioning, and may have contained little equipment aside from an autoclave.

In addition to the declassified documentation found in Kandahar, information obtained through the interrogations of K. S. Mohammed, Hambali, and possibly some others at Guantanamo, Cuba, there was yet one more source of information that was obtained from within al-Qaida regarding its interest in *biological weapons*. Additional fragments of information found at the end of 2001 on computer discs that appear to have belonged to Dr. Ayman al-Zawahiri provide little confidence in the competence of the al-Qaida group to carry out either *chemical* or *biological agent* production. If anything, they detract from assumptions of capability, although they apparently date from early in their efforts. The initial program investment was either $2,000 or $2,000 to $4,000, and, after several months, Dr. al- Zawahiri considered it to have been "*wasted effort and money*."

The group at that time seemed quite constrained in economic resources as well as in applicable talent, and did not at all appear to have the kinds of financial resources available to the Japanese Aum Shinrikyo group. This early effort appears to have been intended to produce a "*nerve gas*" from a commercial agricultural insecticide. Perhaps the most important information in an al Zawahiri memorandum of April 15, 1999, is contained in the following sentences: ...we only became aware of *biological weapons* when the enemy drew our attention to them by repeatedly expressing concerns that they can be produced simply with easily available materials.... I would like to emphasize what we previously discussed—that looking for a specialist is the fastest, safest, and cheapest way [to embark on a biological and chemical weapons program.

Other information indicates that Al-Zawahiri's remark about "the enemy drew our attention to them" refers to U.S. Secretary of Defense

William Cohen's November 1997 national television appearance, which included greatly exaggerated prediction of what his 5-pound bag of sugar standing in for anthrax could achieve if dispersed over Washington, DC.

Early in March 2005, a press item returned to the material found on a computer disc when K.S. Mohammed was captured, and the sentence quoted earlier from the 2003 *Washington Post* story: "... [al-Qaida] obtained the materials required to manufacture two biological toxins—*botulinum* and *salmonella*—and the chemical poison *cyanide*. They are also close to a feasible production plan for *anthrax*." This information was attributed to "U.S. intelligence services quoted in the U.S. media." This can only refer to the general and ambiguous statement by senior U.S. intelligence officials quoted previously. As indicated, the cyanide information dates back to 1998. "Managed to obtain material necessary to make" does not indicate that anything was made, or even that all materials necessary—such as the pathogen—had been obtained. The very next day, an editorial in the *Washington Post* stated, "This country has already experienced one anthrax attack. Security officials have repeatedly stated their belief that al-Qaida and others continue to search for more *lethal bioweapons*." The "others" referred to by Tenet in February 2004 claimed their interest in "CBRN" not specifically "bioweapons." And there are no identifiable public statements by U.S. "security officials" saying that "al-Qaida and others" were searching "for more lethal bioweapons" than anthrax.

It is possible that the "security officials" that *Washington Post* editors had in mind came from another press comment during a March 1, 2005, Interpol conference: "Security officials have long worried of the risk of an al-Qaida attack using biological weapons such as *anthrax*, *ricin*, *botulinum toxin*, *smallpox*, *plague*, or *Ebola*." There are no identifiable statements by any "*security official*" warning of the potential use by al-Qaida of smallpox or Ebola, and the suggestion is highly implausible. On March 31, 2005, the *Report of the Commission on the Intelligence* Capabilities of the United States Regarding Weapons of Mass Destruction became available. The information that it contained concerning the status of al-Qaida and BW was vastly different from the brief paragraph in the report of the 9/11 Commission.

The Intelligence Community concluded that at the time of the commencement of the war in Afghanistan, al-Qaida's *biological weapons* program was both more advanced and more sophisticated than analysts had previously assessed....al-Qaida's biological program was further along, particularly with regard to Agent X, than prewar intelligence indicated. The program was extensive, well-organized, and operated for 2 years before September 11, but intelligence insights into the program were limited. The program involved several sites in Afghanistan. Two of these sites contained commercial equipment and were operated by individuals with special training. Documents found indicated that while al-Qaida's primary interest was Agent X, the group had considered acquiring a variety of other biological agents. The documents obtained at the training camp included scientific articles and handwritten notes pertaining to Agent X.

Reporting supports the hypothesis that al-Qaida had acquired several biological agents possibly as early as 1999, and had the necessary equipment to enable limited, basic production of Agent X. Other reporting indicates that al-Qaida had succeeded in isolating cultures of Agent X. Nevertheless, outstanding questions remain about the extent of biological research and development in pre-war Afghanistan, including about the reliability of the reporting described above.

The sources for the Commission's remarks all refer to classified reports. The two Presidential Commissions appear to have used different procedures to obtain documents for examination, and one suggestion is that, as a result, they did not, in all cases, obtain the same documentation for review. Of course, if the hard information prior to the coalition invasion of Afghanistan regarding al-Qaida and BW was nil, then their efforts will certainly appear "more advanced and more sophisticated" than was previously understood once the actual information in the now declassified documents was found. Is it possible to make any further guesses regarding the substance of the Commission's phrasing?

The information in the "documents obtained at the training camp" and "the handwritten notes" have just been explained in detail in the preceding pages, and concern *anthrax*, *botulinum toxin*, *plague*, and *Hepatitis* A and C. *Hepatitis viruses* are extremely difficult to work with, even for professional virologists. They require cell culture technology, but could in theory be used to contaminate food and water. The most recent U.S. intelligence statements quoted earlier refer more often to botulinum toxin than they do to anthrax. Nevertheless, "Agent X" almost certainly refers to anthrax, with botulinum toxin the most plausible second guess. The key question regarding the information quoted above is whether there is additional documentary or material evidence to support it beyond that already obtained in the papers found in November 2001 and the locations occupied at that time. Those did not indicate success "in isolating cultures of Agent X." And only the Sterne vaccine strain had been available to the group in Afghanistan. The statement that much of the Commission's brief summation was a "*hypothesis*" dependent on "*reporting*," at the same time as there remain "outstanding questions...including about the reliability of the reporting described above," seems to leave much of it an open question and possibly adds nothing of substance to what was already known from the declassified documents. Nevertheless, the reference to two sites containing "*commercial equipment*" may suggest additional al-Qaida efforts beyond those disclosed in the declassified documents and related information so far available to the author.

A member of the Presidential Intelligence Commission was asked if he thought that one should be any less skeptical regarding the intelligence concerning al-Qaida BW activities in Afghanistan than the allegations that had been made by the U.S. administration regarding the status of Iraq's BW program in the years between 1995 and 2002. He replied that one should not be any less skeptical regarding the intelligence about al-Qaida's BW capabilities.

Pakistani press reports indicate that the Pakistani microbiologist who had assisted al-Qaida in information gathering, as well as an alleged "*Yemeni*...studying microbiology at the University of Karachi" were arrested in October 2001. If this is the case, with the facility overrun by U.S. forces in December 2001, Sufaat also arrested in December 2001, and the two above individuals arrested in October 2001, even before US/UK forces had found the documentation in Kandahar, it would appear that the al-Qaida BW program may have been dismantled at the end of 2001. The location of Dr. Khan, in whose home K. S. Mohammed was captured, is known to the Pakistani government. He is incapacitated and is therefore reportedly not a factor of concern. Whether other dispersed individuals of the group have any available work sites for microbiology in any other location is unknown. The statement by French Interior Minister Dominique de Villepin at Interpol's First World Conference on Bioterrorism in Lyon, France, on March 2, 2005, claiming that "...after al-Qaida groups were smashed in Afghanistan, international terrorist groups were still working on chemical and germ weapons in Georgia's Pankisi Gorge" seems highly implausible insofar as it refers to "*germ weapons*." Al-Qaida elements in the Pankisi Gorge had reportedly been killed, captured, or dispersed by a joint U.S.-Georgian special operation in 2002. In contrast to the

French statement, Major General Eric Olson, the second ranking U.S. military officer in Afghanistan, told the Associated Press on February 25, 2005, that he had no indications that al-Qaida was attempting to obtain *nuclear* or *biological weapons*, that there was "no evidence that they're trying to acquire a terrorist weapon of that type."

The three terrorist groups that have been innovative in their methods have one aspect in common: the Tamil Elam in Sri Lanka, the Japanese Aum Shinrikyo, and the al-Qaida organization have all actively recruited among educated, college graduates, and specifically sought individuals with particular knowledge and training. (The Tamil Elam showed no interest in BW, and in only one anomalous incident used industrial canisters of chlorine gas.) K. S. Mohammed completed a degree at a U.S. college, as did Yazid Sufaat. An unclassified summary of information on detainees at Guantanamo states that "More than 10 percent of the detainees possess college degrees or obtained other higher education, often at Western colleges, many in the United States. Among these educated detainees are medical doctors, airplane pilots, aviation specialists, engineers, divers, translators, and lawyers." At least one holds a degree in electrical engineering, another holds a graduate degree in aviation management, and a third holds a masters degree in petroleum engineering. Such recruiting patterns do not automatically translate into either interest or capability in BW, but they would be a key advantage should the interests of such a group turn in that direction, as Dr. al Zawahiri's memorandum quoted earlier indicates.

The reports of the UN Special Commission (UNSCOM) and that of the U.S. CIA's Iraq Survey Group (ISG) provide valuable insights into what one might expect of the initial efforts by a terrorist group. The report of the ISG describes the initial *failures* of the Iraqi national CW program to produce a chemical weapon between 1971 and 1978. The information provided does not clarify if the impediments were in the industrial synthesis of chemical agents or in their weaponization. Nevertheless, this involved a state program, state resources, and a period of 8 years. It was previously known from UNSCOM reports that the same failure occurred in the Iraqi BW program in the "early 1970's," and again between 1974 and 1978. The ISG report also provides information on two efforts made by insurgent groups inside Iraq in 2002-03 to produce very basic chemical weapon agents. Although the services of several chemists were obtained to produce the agents, both these efforts failed. It is uniformly assumed that the production of classical chemical weapon agents as well as their dissemination is simpler than that of classical biological weapon agents.

All of the preceding explains the crucial significance of establishing precisely who was the perpetrator of the 2001 anthrax incidents in the United States, and how and where the anthrax preparation was produced. It is the sole outlier event. Without it, and except for the Rajneesh *salmonella incident*, there would still be no evidence of capability on the part of a nonstate actor to produce a biological agent. There has also been no evidence to date of the provision of assistance by a state to a nonstate actor to produce biological agents. If it should turn out, as is currently assumed, that the Amerithrax perpetrator came from within the U.S. Government's own biodefense program, with access to strains, laboratories, people, and knowledge, then all the conceptions about the significance of the events get substantially altered. It does not alter the fact that it can be done, and that the preparation could have been dispersed in a much more harmful way. But it does affect the crucial question of "By Whom"?—and the projections imputed to traditional "*terrorist*" groups.

With this exception, all attempts by other groups have either failed or been limited to relatively low levels of competence. Reports in 2002 and again in mid-2004 indicated that the investigations into the source of the anthrax used in the U.S. events were yielding results. These had apparently reached the stage that suggested which U.S. laboratory had been the source of the strain used. Nevertheless, there has been no identification of the perpetrator or resolution of the case. In the words of former Deputy Assistant to the Secretary of Defense for Chemical and Biological Defense, Dr. Anna Johnson-Winegar, "We do not know who was responsible. We do not know the source of the anthrax spores (i.e., were they produced, stolen, or purchased?). We do not know the motive for the attacks. And, therefore, we are unable to make intelligent assessments about the likelihood of similar attacks in the future."

In the midst of all these developments, and the information derived from investigations in Afghanistan, October 2003 saw the amazing report that the U.S. DoD had been selling surplus equipment of the kind that could be used precisely for producing BW pathogens, and that some of the equipment purchased by middlemen in the United States had been resold to buyers in, among other countries, the Philippines, Malaysia, Egypt, Canada, Dubai, and the United Arab Emirates (UAE), "for transit to other countries prohibited from receiving exports of trade security controlled items." U.S. officials in the past had identified individuals in Canada, the Philippines, Dubai, and in the UAE who are known to be involved in transshipments to terrorist-supporting countries. The sales had been made by the Defense Reutilization and Marketing Service, which in 3 1/2 years had sold 18 safety cabinets, 199 incubators, 521 centrifuges, 65 evaporators, and 286,000 full-body protective suits.

One can compare this to the reports of the few pieces of elementary equipment found in the al-Qaida site in Afghanistan, and the significance that was given to those finds. The items sold are all available on the commercial market in the United States as well as elsewhere, but DoD's selling price was additionally "*pennies on the dollar*" of the cost of the original items. More accurately, it averaged 10 cents on the dollar of original costs, even if the equipment was unused. In the words of a Congressional subcommittee chairman, "DoD should not be a discount outlet for *bioterrorism equipment*." Export of these items would routinely require an export license if they had been sold for export, and the new Department of Homeland Security (DHS) had already begun monitoring the import and export of the same kinds of materials. Even recent internal DoD regulations had been violated.

The U.S. GAO, the investigative and oversight body that serves the U.S. Congress, was responsible for the report, which documented this absent-minded but gross evasion by a branch of the U.S. DoD of common sense and of existing export controls on materials that could be used for the production of biological agents. After the years of strident alarms regarding the interest of international terrorist groups in *biological weapons*, and after the U.S. domestic anthrax incidents in 2001 and the enormous post-9/11 buildup of "*Homeland Security*" to protect against bioterrorism involving expenditures of $7-8 billion by the time of the GAO report in the fall of 2003, this was an amazing demonstration of one dysfunctional branch of government facilitating exactly what the rest of the government was ostensibly trying to protect against. To add to the irony, the 3 1/2-year period investigated--between October 1, 1999, and March 31, 2003—places its onset during the tenure of none other than Secretary of Defense William Cohen, avid warner of "the *bioterrorism threat*."

12

RARE STRATEGIES

To an extent, this was a legitimate concern. The heart of a nuclear fission device (bomb or reactor) is a quantity of fissionable material such as U235. U235 is an *isotope* of *uranium* that constitutes about 1% of natural uranium (most of the rest being U238, which is not fissionable). So even if you have a chunk of pure uranium, you need to separate the U235 out in order to have something useful for a bomb.

It is much harder to separate one *isotope* of an element from another than it is to separate two different elements or chemical compounds. Typically, different elements will have different chemical or physical properties. So you put your mixture—ore, for example—through some process that has a different effect on one than the other. A simple example is using a still to separate alcohol from water. Alchohol boils at a lower temperature than water, so if you heat the mixture to somewhere between the two temperatures, the alcohol boils off to be collected in your condenser, while the water stay behind.

But with isotopes, the properties are so similar that it is virtually impossible to distinguish atoms of one from the other. Uranium itself is a very heavy, fairly inert metal. Boiling it is out of the question. To separate U235 for the original bombs during World War II, the United States built the Oak Ridge plant, a city-sized processing facility that still took months to produce enough U235 for one bomb.

The bottom line is that isotopic separation is one of the major technical challenges that forms a serious barrier to entry into the nuclear weapons club. And that's a good thing. Now it turns out that if you only have a few atoms of uranium, or anything else, it is not all that hard to separate them. The difference between U235 and U238 is about 1% in weight. Ionize your atoms and run them through a mass spectrometer, which is an instrument that fires them through a magnetic field. The field makes them curve. The lighter ones curve more easily, the heavy ones don't curve as much; so they arrive at different spots on the other side of the field and are thus separated. The mass spectrometer is a scientific instrument that is very useful in analyzing tiny samples of a substance, but it is of no use for the kind of quantities you'd need for a bomb.

With *nanotechnology*, you don't even need to fool with the tricks like ionization and so forth that bulk-technology mass spectrometers use. Essentially you just build a gadget that grabs an atom and weighs it. If its atomic mass is 235, it goes in bin A, and if it is 238, bin B. Gravity is of no use for weighing atoms, but a centrifuge would work fine. Or more precisely, you build a million billion such gadgets, together with the machinery to distribute atoms to them and put the results back together. This composes a machine that sucks in 5,000 pounds of chemically pure uranium dust on one end, dumps 4,950 pounds of depleted uranium dust out the bottom, and produces a 50-pound sphere of U235 at the business end. Woops, better make that several pieces that could assemble to make a 50-pound sphere, and make sure you keep them apart.

Self-replicating Weapons

There is a notion that nanotech weapons will be bacteria-like replicators that can live off the land of the enemy's territory and ultimately reduce it to dust or have any other devastating effect that the weapon-wielder desires. There are two points that need to he stressed about such a scenario: First, the weapon scenario is much more likely than that of some kind of accidental replicator release. The difference is that commercial nanomachines will be made for high efficiency (and profitability), and thus be engineered to run fast, from high-energy foodstocks, be controllable, and produce useful outputs. Trying to live off the natural world violates all these desiderata. Accidentally building a nanomachine that would operate in the wild would be about as likely as accidentally building a car that sawed down trees, cut and split them into firewood, and stoked a boiler firebox with them.

It would be possible to design such machines on purpose, however, and weapons have always been the application where efficiency and other such considerations are last on the list. If someone designs a Nanomachine capable of operating in the wild, it will almost certainly have been done as a weapon. The second point, however, is that while it would have possible to design army tanks that burned wood and thus could a in the forest, no one ever did. In fact, actual weapons system strongly in the opposite direction. The reason is that although efficiency is at the bottom of the list of design criteria for war machine power and speed are at the top, and stopping to eat grass or whatever is not the way to get them. Commercial nanomachines will mature more of themselves from high-purity; high-energy raw material and so will military ones.

There just is not the energy available for the taking in the natural world to power machines at the level our current-day machines operate, much less nanomachines. Wood-burning machines cannot even compete with coal-burning ones, much less oil- or hydrogen-burning ones. Although it is too far out and too technically uncertain to mention in other connections, it would not be surprised to see nuclear-powered military machines (radioisotope, not fission). Note that the military currently uses (fission) nuclear-powered ships, which are not cost-effective for commercial use. Where today we have intelligent bombs and missiles, nanotechnology could give us smart bullets. Adding AI in nanocomputers to the mechanical capabilities would give us robot soldiers that could fold up to the size of a rat to creep through cover unfold to the size of a human to manipulate objects or weapons and fly at: hundreds of miles an hour. They could be virtually invisible using phased-array optics skins. Their motions could be too fast for a human eye to track. To be prepared for war, a nation need only have a widespread nanomanu-facturing base that could produce such robots, with matching vehicles and weapons, should the need arise. When push came to shove, the actual army could be produced in under an hour.

A good-sized patch of phased-array optics could focus light on a target. If the roofs of buildings were covered with these, they could act as solar collectors, radars and optical cameras, spotlights, and if tilted enough, streetlights. If necessary, and if used in a coordinated way from many buildings at once, they would function as laser like antiaircraft artillery, In a post-9/11 world it is regrettably understand able why that might be desired. Private spacecraft, in particular, would be viewed with concern, and might be allowed only if there were some "instant-on" distributed air defense.

WAR

It the twentieth century, the cause of human disaster that clearly stood out above all the others was war. Technology clearly contributed to the deadliness. Will nanotechnology make war worse, or more likely? After all, a nano weapon can't kill you any deader than a hydrogen bomb. What's more, since the development of nuclear tipped ICBMs, a remarkable thing has happened. Nations have quit going to war against other nations that are armed with them. The author is of the opinion that this has a lot to do with the act that such weapons put the political leaders of the attacking country in direct physical danger. As Ambrose Bierce quipped, the meaning of *rear* in American military matters is that part of the army closest to Congress.' Nanotech weapons will only exacerbate this phenomenon making it easier to target the politicians. If so, war could suffer from a further decline in popularity.

Another phenomenon to consider is the industrial revolution and the *Pax Britannica*. If one nation or aligned group does attain a level of technological dominance comparable to Britain in the nineteenth century a similarly peaceful period might ensue. At the moment, the United States and Europe account for a large majority of known nanotech efforts. Although certainly not perfect, the western democracies would probably make a reasonably enlightened and stable world hegemony. The one scenario that seems to merit some concern is the opposite—where other countries develop nanotech first. The United States might then engage in some serious saber-rattling rather than slip quietly behind. Here is a vignette that, thankfully, is completely imaginary at the movement:

It's 2015. In the United States, business as usual has been allowed to prevail. Interest in science has continued to decline. Virtually all the scientists and engineers our universities produce come from, and most return to, other countries. Funding for research is mostly for medical applications, and that is mired in political debates over stem cells and choked with red tape attempting to make it totally safe.

Meanwhile, China has pushed ahead on a broad range of fronts and has produced Stage III replicators. Products begin to appear from China that cannot be made economically anywhere else. No official notice is taken in the United States because our labs can still produce better stuff in expensive, one-off; form. The Chinese are accused of "dumping" and some nanotech products are banned.

China proceeds to Stage IV and Western technology begins to look distinctly second-rate. They are rumored to have engineering design supercomputers. The latest generation of Chinese jets and spacecraft has significantly better capabilities than ours and was designed and produced in half the time. The US military sounds an alarm.

The administration undertakes a crash program to demonize nanotech as "weapons of mass destruction" and get a UN resolution prohibiting it anywhere in the world. This stalls in debate and goes nowhere. The United States makes a unilateral ultimatum to China demanding a halt to all nanotechnology China demurs, and announces that if attacked with nuclear weapons, it will release aerovores into the atmosphere. It is perhaps worth noting that during the industrial revolution, France engaged in a series of wars—that were ultimately won by England at Waterloo.

Terrorism

On September 11, 2001, in a limousine on the way to Newark airport was observed. Many persons had seen, watching television at home, a plane flown into the World Trade Center. The World Trade Center looked like a pair of smokestacks. The question of terrorism can be divided into two parts: why do people try it, and how do they succeed? The first is a constant of human nature: take people and make them feel powerless, put upon, exploited, and having nothing to lose. Religion is often involved but any strong ideology will do. Some of them will explode in whatever way they can. It is the nature of the beast. The second is more complicated, depending on technology and other circumstances.

If instead of concentrated, regulated, commercial air travel, all the passengers flying that day had been in private air cars, September 11 would have amounted to a handful of carjackings, if that. Personal air craft the size of cars simply could not have brought the towers down. Al Qaeda did not field enough operatives to destroy the towers on a piecemeal basis. You may remember they had already tried explosives-filled vans some years before.

That is not to say that private aircars would have been economically feasible in 2001, but to point out that decentralization is one of the best defenses against terrorism. A terrorist is at his most advantageous position when the defense has made a whole population helpless in hopes of disarming whatever terrorists may be among them. Public transportation is a favourite terrorist target: airliners in the United States, subways in Tokyo, buses in Jerusalem, trains in Madrid.

Oppression, or perceived oppression, generates the kind of anger, shared by a whole community, sufficient to marshal significant resources and motivate suicide agents. In the short run, this is purely a matter for statesmanship. In the long run, the kind of independence nanotechnology can provide may help. In the medium term, the problem of making the transition is delicate and difficult, but must be faced; because in the long term, some terrorists will get nanotech weapons anyway. If nanotechnology is a widely based main stream development, the effects are not likely to be worse than what the people have seen. If nanotech is decentralizing capabilities are used, terrorism could get harder to pull off. It would also be easier for the mainstream to retaliate. As of this writing, after more than two years and after conquering two countries on the other side of the world, American forces still have not captured Osama bin Laden, despite a $50 million reward. Some of the more useful drones: *semirobotic aircraft*. With nanotechnology we could hood the country with drones the size of flies. Hiding would be much more in ore difficult.

Current-day drones have been equipped with air-to-surface missiles and used to strike buildings and vehicles. *Nanodrones* could carry stings of various kinds and target individuals for lethal, debilitating, or punitively painful doses of chemical or biological agents.

A final note on terrorism: In my opinion, the major high-tech terrorism threat is not nanotech, but biotech. Not only have not we caught bin Laden, we have not caught the anthrax mailer, either. He or she seems likely to have been an American working alone. Alone!—while bin Laden and the Taliban had thousands of followers and supporters.

The *nanotech version* of a bacterium, a loose mechanical replicator of microscopic size, might be more efficient than a natural one. But there can be no nanotech version of a virus. A bacterium carries its own building machinery, which nanotech can make in a more efficient form. But the virus is even better—it carries no machinery at all. It uses the machinery in your own cells for its purposes. It consists of, in simple terms, just the genes needed to make copies of itself, and your cells do all the work. That is a level of finesse nanotech could not use. But note that viruses could be made right now with DNA you can order online—$2.35 per base pair—and with equipment in a high school bio lab.

As far as technological timescales are concerned, *biotech* is already here. It is as easy or easier to hide than nanotech. And people have not come close to realizing the scope of threats it could produce. A properly tailored superflu could kill millions, given, say, twenty suicide circle agents to start the spread all across the country at the same time. Or someone could develop a mad cow prion for chickens. Or a Cipro-resistant anthrax. Or transplant the botulin toxin gene into algae and drop samples into reservoirs. Although these are not nanotech threats, nanotechnology (nanomedicine in particular) would be invaluable to help fight them.

Economic Status

In the foregoing, we have addressed the applications of nanotechnology as if it were the United States and/or Europe that developed and deployed it. However, that is far from a certain outcome. Although we have a great head start at the moment, we could easily wind up as the Scotland of golf, the England of tennis, or the France of the industrial evolution. The reasons, as explained earlier, are manifold: bureaucratic business as usual here; a decline in the number of Americans going into science and technology juxtaposed with a huge increase in the number of science and technology students in Asia; and, probably, a greater willingness to take risks there.

Add to that, of course, the fact that Asia is not monolithic, but has numerous competing nations. This has several effects: competition is a strong motivator and acts as a spur to prevent bureaucracies falling into the lassitude which is their normal state. Look at what NASA achieved in the 1960s, when there was a "space race," compared to what it has done since. Second, in a project of research, one country might commit to one approach, and another, knowing it'd be behind if it just followed suit, could try something else that might turn out to work better. But the advantage is more subtle than simply trying different things.

In a hierarchical organization, when a contrarian scheme begins to look like it might have been a better idea than the reigning orthodoxy, there's a strong temptation on the part of the higher-ups to quash it. In a situation where the competing elements are separate nations, this can be avoided to some extent.

If nanotechnology were developed in Asia with a significant lead time, the United States would be reduced to the status of a second-rate power. We are used to living behind the buffers of the great oceans and projecting our power where we wish on the globe. We could lose both—the dominant power might simply decide to tailor the Earth's climate to suit itself, for example, without bothering to ask us. You can be sure that the economic realities would shift, and we would be on the wrong end of some bad deals, shut out of monopolies, and, in general, have a lower standard of living and fewer opportunities than if we maintained the edge.

NANOPARTICLES

One danger that has been mentioned in the press in connection with nanotechnology is the release or biological effect of *nanoparticulates*. Today's nanoscale technology is producing some new products, such as paint that resists bacterial growth. But the effects of such substances, while different in detail from current substances we deal with do not differ in any significant overall way. After all, ordinary untreated wood contains substances that inhibit bacterial growth. Ordinary chemicals are magic of molecules even smaller than *nanoparticles*.

The big difference with real, eutactic, nanotechnology is in the opposite direction. Current technology cannot produce products and energy without releasing floods of molecular-level junk into the environment. If you want to produce billions of chemically reactive nanoparticles, just build a fire in your fireplace, or drive your car, or cook something. If you can smell it, it is releasing molecules that fill the air.

Eutactic processes change that. Every molecule is accounted for. Rather than dump odd chemicals and particles into the environment, they can be broken down to small constituents and used to build the next thing. Instead of producing carbon dioxide, for example, the nanofactory would release pure oxygen, and save the carbon to make diamond. The only combustion product normally released would be water.

OTHER DANGERS

Suppose you were transported back to 1900 and asked to explain to people what dangers were posed by the motor car. If you said pollution. You had be laughed at. You will understand why if you have ever ridden a train with a coal-burning steam locomotive. Steam engines of the day produced a lot more noxious smoke than gasoline engines. A report of deadly crashes might find more favour—people were scared of these unusual mechanic monstrosities. But you'd still be wrong: cars have saved a lot more lives than they have taken, and the average life-span has increased by more than twenty years since 1900. Traffic jams? Hard to imagine in a country with their acres per person. Fragmentation of the traditional extended family and dissolution of sexual mores? Excuse me, sir, are we speaking the same language?

The fact is that if you had to wait to die in a car crash, you had live in average of 6,700 years. Traffic jams are a nuisance, but except in the most densely populated areas, there is no great demand for alternatives such as public transportation. In my own experience the greatest problem with cars in cities is parking.

The social effects are perhaps somewhat closer to the mark. In fact, the twentieth century has seen changes in our life-styles that would have horrified our great-grandfathers. But are we the worse for them? In some cases, perhaps; in others, probably not. In most cases there are compensations—the changes have made things different rather than particular better or worse. What is more, the changes came about from a confluence of causes, including cars, two world wars, radio and TV, the pill, the jet airliner, the computer, and the Internet.

The printing press has been blamed for the Reformation and some hundreds of years of religious wars. This could obviously have been handled better, but these were the growing pains for moving from a largely illiterate culture to a literate one. It does give us a clue as to the real dangers that lie ahead, though. Entrenched power will react strongly to change, even if the change ultimately benefit most people.

Imagine the politician or corporate or labour leader, who became possessed of a device that would allow him to put you in a chair, clap a helmet over your head, and five minutes later have you step out, a fanatical devotee of his, anxious to grab others and hold them down in the same chair. This is not something that anyone has any idea how to build right now, even given full-fledged nonotechnology. But biological knowledge is advancing as fast as any other realm of science, and faster than most. I had find it surprising if by the end of the century; something like this were not possible.

Suppose a great nation decides to do the same with its army, promising to set them back to normal after the "emergency" is over.

Suppose every citizen is put through a milder form of reprogram-flung "to *prevent terrorism.*"

Nanodrones, described above as an *antiterrorism weapon*, could easily be misused as an assassination weapon. Nanodrones are early, not advanced nanotech, probably Stage III or thereabouts. Crude, only by comparison, microdrones can be and are built today. These have cameras only; as far as the author is aware. Even they threaten significant social change following to ubiquitous surveillance.

Being constantly spied on is one thing. Being subject to termination with extreme prejudice on someone is whim is another. Nanodrones could be quite difficult to stop, and virtually impossible to trace. If they became generally available, it might be a bad time to be a lawyer. More seriously, note that the United States already has used its (full-sized) drones for assassination, and that governments as a class, over the twentieth century; killed tens of millions of their own citizens and at least as many of the citizens of other countries. Even if governments can be forced to open their operations to public scrutiny and oversight, other organizations will make obtaining nanodrones a high priority. It will be a bad time to be Jimmy Hoffa.

Note that like the other threats, nanodrones are not nearly so formidable to a population that is equipped with nanotechnology. A drone would be useless against someone living in a nanoskin or surrounded by Utility Fog, for example. The danger is that governments might, in a protectionist panic, limit access to countermeasures. This could easily be a slippery slope to a situation with nanotech-equipped officials and powerless, unprotected commoners. And government of the people, by the people, for the people would perish from the Earth.

Living in Interesting Times

The real dangers that will come with nonotechnology are not the end of things that play well on a movie screen or can be explained in sound bites. They are not dangers from the technology itself, but the effects of shortsightness and greed in the face of a revolution in human affairs. Every major, life-changing development has been accompanied by some strife. Much of the strife has been, in retrospect, unnecessary; but it never seemed that way at the time. In many cases, as in the Reformation, people didn't really understand what they were fighting over until long after the dust had settled.

One of the best ways to prevent, or at least minimize, strife as nanotechnology is developed is for there to be a broad understanding of what benefits it can bring and what the dangers really are. If nanotech remains the dimly understood magic of a powerful few, trouble lies ahead. But if people widely understand how personal manufacturing technology can bring independence and a comfortable life-style to everyone, and the way is made clear for this to happen, cooperation—and sanity—may just prevail.

13

BIOLOGICAL TERRORISM

Preparing for *biological terrorism* has more in common with confronting emerging diseases than with preparing for *chemical* or *nuclear attacks*. Biological terrorism will bypass the quick-response teams that would be critical to coping with attacks using chemical or *radioactive materials*. Unless a biological attack is announced, or discovered while still underway, it would not become clear that something was wrong until victims began to show up in doctors' offices and hospital emergency rooms. The disease agents likely to be used as terrorist weapons may incubate for hours, days, or even weeks before their victims feel any symptoms. More than likely, the earliest symptoms will mimic those of a bad cold or flu. Sufficiently subtle or distributed terrorist attacks may, at least initially, be indistinguishable from naturally occurring infections or outbreaks. This emphasizes the importance of training physicians and other health care workers to determine rapidly, on the basis of the earliest cases, that an unusual infection is involved. Such training is currently still largely missing.

Because of the *incubation delays*, no nation can protect itself by simply screening travelers at its borders. Nor can a country such as the United States hope to inspect more than a small fraction of the food it imports daily. As agricultural markets become increasingly global, the potential vulnerability of nations to food-borne natural or intentional disease will continue to increase. Protection against both emerging diseases and biological terrorism must instead rely on *disease surveillance*.

The synergy between the responses needed to meet these two threats is a powerful one and the United States and global organizations should take full advantage of it. Improving public health surveillance for biological terrorism must have both strong domestic and international components. It also requires better coordination between public health, law enforcement, and intelligence agencies. How do we measure success in these endeavors? The success of a strategy of prevention is invevitably difficult to prove unambiguously. The more successful a prevention strategy proves to be, the more a metric for success will need to measure surveillance capabilities rather than response. The absence of an undesired, perhaps catastrophic, outcome can never be proven to be due to any particular level of preparedness. Given the importance of prevention, policymakers must become comfortable with this less direct metric of capabilities. Yet it is, after all, a familiar one in the national security realm—as familiar as the *Cold War* strategy of deterrence.

Domestic Measures

Because incubation delay periods for many diseases are longer than international flight travel times, we cannot hope to stop all diseases at the borders of the United States. Nevertheless, screening and quarantine efforts at ports of entry and inspection of food imports provide an important component of public health surveillance. In 1995, the Committee on International Science, Engineering, and Technology (CISET) of the Clinton Administration's National Science and Technology Council (NSTC) called for the strengthening of screening and quarantine efforts at ports of entry into the United States. With respect to food safety, the Administration issued the *National Food Safety Initiative*, which includes improved coverage for imported foods (as well as for domestic produce, seafood, and livestock). Recognizing the importance of acting abroad to ensure domestic protection, the Initiative calls for the Food Safety and Inspection Service (FSIS, within the U.S. Department of Agriculture) to provide technical assistance to countries whose products are implicated in food-borne illnesses. These initiatives should improve surveillance for both natural and artificial outbreaks.

Further improving domestic surveillance requires improving sensitivity and connectivity along the chain from physicians to national health authorities. We will consider each link in that chain in turn. First note, however, that it may be unrealistic to expect domestic public health agencies to find substantial resources to improve surveillance for biological terrorism within their existing budgets. For example, the formal mission of the CDC is "To promote health and quality of life by preventing and controlling disease, injury, and disability"; its mandate is to provide the greatest good for public health. It is inevitably difficult to draw resources from programs that are protecting the lives of Americans from day-to-day life-threatening illnesses and redirect them to surveillance for future attacks that may or may not ever take place.

Expanding public health surveillance explicitly to include surveillance for biological terrorism will require new resources. An announced biological attack, or one discovered while under way, will require first responders who are appropriately trained. Other biological attacks must first be recognized by pathologists, physicians, and other health-care personnel in family practices, clinics, and hospitals. It is important, therefore, that these medical professionals have some knowledge of the clinical presentations of likely biological terror agents. There is a broad parallel with actions recommended in the Clinton Administration's CISET report for addressing emerging diseases. That document called for expanded formal training and outreach for health-care providers. The National Institutes of Health (NIH) and CDC have responded by writing to medical and microbiology associations and other professional organizations urging them to focus training and certification programs on emerging diseases, and continue to sponsor meetings on related training needs.

Similar actions on the part of NIH and CDC to raise physicians' awareness of biological agents should be undertaken. A first step is the article "Clinical recognition and management of patients exposed to biological warfare agents" in the August 6, 1997, issue of the *Journal of the American Medical Association*. However, the best way to ensure that busy physicians improve their expertise in this area is to require relevant knowledge in medical school curricula and certification examinations, and to offer appropriate training.

In FY97, Congress appropriated $52.6 million to the Department of Defense (DoD) to implement various domestic preparedness programs. These funds were used in a variety of ways, including the development of a Chemical-Biological Rapid Response Team (CBRRT) and to procure additional equipment for the U.S. Marine Corps Chemical Biological Incident Response Force (CBIRF). The DoD also began to train trainers in 120 U.S. cities to prepare for and respond to emergencies involving weapons of mass destruction. By the end of 1997, twenty-seven cities had received visits. The DoD expects to discontinue this training after FY99. Lead agency responsibility for this training should therefore be transferred to the *Public Health Service* (PHS). The

Federal budget currently devotes some $7 billion annually to unclassified terrorism-related programs. It is critical that within this vast budget, sufficient and ongoing resources be found to train the local physicians and other first responders to any likely biological attack.

In June 1998, President Clinton requested an additional $294 million from Congress to deter and respond to terrorist incidents involving biological and chemical weapons. This request included continued funding for local training programs. The first step in improving sensitivity for incidents of biological terrorism is for the federal government to make this a long-term, sustained commitment to training for the nation's physicians, pathologists, and other first responders.

Next, regional centers of excellence, building directly on the best state public health laboratories, should be established with the capability for rapid diagnoses of clinical samples from within their geographic areas. These regional centers must have the trained personnel and diagnostic tools necessary to accomplish this mission, and connections to both local and national institutions must be assured.

To improve the ability to identify rapidly and accurately the early stages of a possible bioterrorist attack, some six to ten regional sites around the United States should be designated for substantial improvement in both epidemiological and rapid diagnostic capabilities. This capability for high-volume rapid diagnostics differs from the traditional expertise of national reference laboratories. These new regional centers of excellence should build directly on the best of the state public health laboratories in order to minimize additional expense. The President's 1998 request to Congress also asks for an additional $43 million to improve the ability of public health centers to recognize and share information on outbreaks of suspicious diseases. This important request, which if implemented would improve both sensitivity and connectivity, should be fully funded.

At present, too little attention is being given to developing rapid diagnostics appropriate to this sort of laboratory setting. For example, state or regional laboratories will need diagnostics for biological agents that are capable of thousands of sequential assays. Such diagnostics would not have to be hand-held, and would not necessarily need to employ cutting-edge technologies. But they would need to be robust and reliable. Similar equipment may also be useful for the Army and Naval Research Facilities overseas; at times of major outbreaks of infectious diseases, these regional reference labs may be swamped by samples requiring examination. (The Department of Defense operates infectious disease laboratories in six countries overseas. These labs conduct epidemiologic investigations, diagnose diseases, and recommend control measures. They conduct research on diseases of mutual interest to both the host country and the United States.)

Whereas high-volume diagnostics might remain unstressed for long periods in the United States, at reference laboratories overseas they would more likely be challenged by use in real outbreaks; this could provide a valuable opportunity to refine these tools under real conditions. The relevant federal agencies should ensure that at least one agency is working to meet needs for robust, rapid, high-volume laboratory diagnostics. A related requirement, also with resource implications, is to maintain a cadre of individuals in the United States with expertise in the diseases likely to be employed by terrorists. For example, at present CDC has the only laboratory in the world that serves as a reference laboratory for plague. There is only one full-time employee at that laboratory with experience and training in plague epidemiology and treatment.

The agencies that would be called upon to perform these tasks in the event of outbreaks of biological warfare agents should complete an inventory of critical personnel needs. It is unlikely that agencies will contribute additional positions to individuals with expertise in diseases rarely encountered in the United States. (In 1996, five cases of plague were reported in the United States, of which two were fatal; both decedents died before plague was diagnosed.) Yet individuals with training appropriate to most biological warfare agents are important for responding to outbreaks abroad, since most biological agents are also naturally occurring diseases in one or another region overseas. Funding and positions should be provided for sufficient individuals to maintain national expertise in those diseases likely to be used for biological terrorism.

International Networks

Surveillance for disease outbreaks overseas must also be improved. The surest way to alleviate human suffering, as well as to prevent disease from reaching America's shores, is to detect and stop outbreaks quickly while they are still abroad. The capacities that are needed—trained health care workers and epidemiologists, regional laboratories with reliable diagnostic equipment, good communications, and the ability to send in teams of experts—will help spot both emerging diseases, as well as any outbreaks resulting from the use, testing, or accidental release of biological agents.

The first step should be to improve the existing international network for the detection of *infectious diseases*. There are currently too many geographic holes in the international disease surveillance system. The *World Health Organization* (WHO), the obvious choice for a multilateral solution, has in the recent past been viewed with skepticism by many experts, due in part to its limited resources. "By the time WHO realized there was an AIDS epidemic it already existed on four continents. That's WHO preparedness and emergency response for you," commented D.A. Henderson, the physician who led WHO's smallpox eradication effort. But there is cause for growing optimism. In 1995, the World Health Assembly, the legislative body of the WHO, adopted a resolution calling on WHO to lead, strengthen and coordinate international efforts to respond to emerging infectious diseases.

As a result, the Division of Emerging and other Communicable Diseases Surveillance and Control (EMC) was established, with a mission to strengthen national and international capacity in the surveillance and control of communicable diseases. The WHO/EMC publishes in both print and electronic formats the bilingual English/ French *Weekly Epidemiological Record* and the electronic *Disease Outbreak News*. It is also compiling a searchable database of the WHO/EMC collaborating centers worldwide and, jointly with the World Bank and the Joint UN Programme on HIV/AIDS (UNAIDS), connecting the collaborating centers electronically.

Simultaneously, the Program to Monitor Emerging Diseases (ProMED), an international non-governmental group of infectious disease experts, has established an electronic reporting system open to unconfirmed reports of disease outbreaks. This system parallels the more strongly filtered WHO *Rumour Outbreak List*. These steps are reminders of how "*connectivity*" increasingly means access to electronic mail and the World Wide Web, and the extent to which international health security is enhanced when all nations, including developing nations, gain access to these networks. This is an area where U.S. agencies such as the U.S. Agency for International Development (USAID) may be especially well placed to provide technical assistance and support, working together with international agencies and host governments.

There are two broad categories into which improvements in international surveillance for bioterrorism may be divided. The first is ongoing "*background*" surveillance with the intention of recognizing outbreaks as they occur, while the second involves a directed response to a specific outbreak that has been detected. These latter cases remain in the category of surveillance as long as the "*response*" includes an investigation whose goal is to identify the nature, extent, and origin of a disease outbreak. By this definition, the CDC team dispatched to the Ebola outbreak in Kikwit, Zaire, in 1995 was involved in surveillance (in addition to its critical missions of providing medical care and containing the outbreak).

Investigations of recognized outbreaks (especially those deemed suspicious) and the ability to identify the responsible organism or strain are critical, but these capabilities are dependent upon a surveillance system operating in the background that is able to detect outbreaks as they occur. While egregious attacks or accidents in biological warfare programs may be difficult to miss, the ability to verify the *Biological and Toxin Weapons Convention* (BWC) and to deter would-be violators is enhanced by having as sensitive a public health surveillance network as possible, one that will detect outbreaks that are subtle or identify less-than-subtle outbreaks in their earliest stages. Moreover, such a network provides the best opportunity to stop an outbreak before it reaches the United States. The surest route to such a capability is to improve the international surveillance system for emerging diseases. In 1996, Vice President Gore announced the Clinton Administration's new policy for responding to emerging infectious diseases. Under that policy, President Clinton directed that the U.S. government would "work with other nations and international organizations to establish a global infectious disease surveillance and response system, based on regional hubs and linked by modern communications technologies."

Such new regional networks should be integrated with the five independent monitoring and alert systems of the WHO/EMC. Information from these systems are made freely available on the World Wide Web and in other fora. One of these systems comprises the WHO Collaborating Centers, a network of over two hundred laboratories and institutions around the world. Collaborating Centers carry out specific activities on behalf of WHO and provide information on disease distribution, while providing laboratory diagnoses and training in the host nation. Host governments agree to allow the Centers to report directly to WHO, without first going through the government. However, there are large regions of the world where these Centers are absent or rare, including Eastern Europe and much of Saharan and sub-Saharan Africa, Central America, and Southeast Asia. The new regional networks would help fill these gaps.

The CDC has thoroughly examined how to support the development of international regional networks of closely linked epidemiology and laboratory programs to promote disease surveillance. These plans were outlined in the CDC document, Addressing Emerging Infectious Disease Threats: A Prevention Strategy for the United States. In 1993, ProMED had endorsed a system similar to that proposed by the CDC. Ten medical centers, strategically located in the developing world, would serve as global health sentinels. The centers would be built directly upon the most capable existing facilities, in order to minimize expense, but would need to be given priority for international assistance. The initial costs for such a network could be modest, perhaps $10 to $20 million per year. This would represent a small fraction of the $7 billion in unclassified terrorism-related programs the U.S. government currently spends. If the United States wishes to improve global surveillance for either emerging infectious diseases or incidents of biological terrorism, taking the lead in developing an international surveillance network is perhaps the most important commitment it could make.

A proposal for substantially augmenting the global monitoring system might also provide a useful tool in the BWC negotiations. Under Article X of the BWC, States that are Parties to the Convention "in a position to do so shall also cooperate in contributing individually or together with other States or international organizations to the further development and application of scientific discoveries in the field of *bacteriology* (biology) for prevention of disease, or for other peaceful purposes." Augmentation of global surveillance for infectious diseases could be presented as cooperation under Article X.

One element of the global monitoring system being strengthened by the Division of Emerging and other Communicable Diseases Surveillance and Control (EMC) provides a model for public health reporting that sidesteps explicit references to biological agents. The International Health Regulations (IHR) are the only international public health legislation that requires mandatory reporting of infectious diseases (*cholera*, *plague*, and *yellow fever*). To transform the IHR into a global alert system, WHO is revising them to broaden their scope to include many diseases for which they currently make no provision.

The approach is to require notification of five specific *clinical syndromes* (respiratory, neurological, antimicrobial resistance, diarrhoeal, and hemorrhagic). Reports of syndromes will be followed by reporting of specific diseases once the diagnosis is known, but action can commence even before a laboratory diagnosis is made.

From the point of view of those concerned with incidents of *biological terrorism*, these five syndromes will capture outbreaks due to biological warfare agents as well as natural causes. This in turn could provide an appropriate way for regional networks to *de facto* participate in surveillance relevant to biological agents without having to do so explicitly.

Biological and Toxin Weapons Convention

A verification regime for the BWC is hampered by the easy availability and dual-use nature of the microbiological technology needed to culture disease organisms. In this light, investigations of unusual or suspicious outbreaks of disease may be the best option for improving verification. The United States should continue to work for the right of the global community under the BWC to investigate suspicious outbreaks wherever they occur. Would-be developers of biological weapons should fear that if an accidental release occurs, the world may discover the resulting outbreak and pinpoint its origin. These same investigations may lead to the identification of unusual but natural outbreaks as well.

The Ad Hoc Group of Governmental Experts (also known as Verification Experts or VEREX) created by the Third Review Conference for the BWC in 1991 explored twenty-one different possible verification measures for the BWC, including surveillance of publications and legislation, scheduled declarations of activities, remote and on-site inspections, and others. The VEREX concluded that no combination of measures could be found that would uncover violations with a high degree of confidence. A consideration of the demands placed by attempted verification of the BWC makes it clear why the task is so difficult. For example, sales estimates of fermenters in the range appropriate for illicit pilot-plant production of biological agents numbered 3,600, in 1984 alone; such plants could be attached to most major universities or biological firms.

Because of such practical considerations, BWC negotiators have narrowed the likely categories of declarable facilities to biosafety level 4 laboratories, those facilities producing vaccines or biopesticides, and military and biodefense programs. Otherwise, the number of sites requiring verification is just too large. Moreover, U.S. companies have concerns regarding the protection of industrial secrets which might be compromised by inspections under a BWC regime. In any case, the experience of the United Nations Special Commission (UNSCOM) in Iraq makes it clear that even comprehensive mandatory declarations and intrusive challenge inspections of a range of biocapable facilities is insufficient to guarantee compliance: Iraq developed and maintained a biological weapons capability while under the direct scrutiny of UN inspectors. Investigations of unusual or suspicious outbreaks of disease may therefore be the best option for improving verification of the BWC. President Clinton endorsed such a measure in his speech to the UN General Assembly in September 1996. In light of the history of accidental releases in biological weapons programs, a right of investigation could provide a deterrent to such programs, and a possibility of detecting violations.

The Defense Special Weapons Agency has conducted a disease outbreak exercise that corroborates the utility of on-site epidemiological investigations in making determinations of the nature of unusual disease outbreaks, and it has outlined preliminary criteria for recognizing possible biological weapons events. Mechanisms for the initiation of formal on-site epidemiological investigations of suspicious disease outbreaks are under discussion in ongoing negotiations for a protocol to the BWC, and the United States should place high priority on these negotiations. Because it is important that health agencies be able to operate overseas in a transparent manner, directed investigations into suspicious outbreaks are probably best left to teams specifically organized under the BWC. A different kind of deterrent stems from some of the same molecular biological technologies that could facilitate the engineering of improved biological agents.

Genetic fingerprinting (or more broadly, biological signatures tracking: the ability to identify, distinguish, and establish relationships between particular strains of organisms through biochemical or molecular biological analyses of those strains, for example via the development of a library of DNA sequences corresponding to different strains of viruses and bacteria) would help assure would-be attackers that even a secret biological release might nevertheless be tracked to its source. Greater transparency, to include the exchange of strains of organisms held in the national laboratories of individual nations, could facilitate this goal. A DNA-sequence database for different strains of organisms, especially for those associated with weapons programs, is very important for investigating either domestic or international outbreaks. Biological signatures tracking and attribution could be a powerful tool for identifying when an outbreak is artificial and who its perpetrator might be. Biological signatures tracking and attribution research and development should receive high priority for continued and additional funding.

Improved Coordination

Coordination between public health and civilian emergency response agencies is improving. Under the National Food Safety Initiative, the four federal agencies charged with responding to outbreaks of food-borne and water-borne illnesses (the Food and Drug Administration [FDA], CDC within Health and Human Services [HHS], the Food Safety and Inspection Service [FSIS] within the U.S. Department of Agriculture [USDA], and the Environmental Protection Agency [EPA]) are establishing the Food-borne Outbreak Response Coordinating Group (FORCG) to develop standardized procedures for the rapid exchange of data and information associated with food-borne illness outbreaks.

The HHS, USDA, and EPA will designate the Assistant Secretary of Health, the Under Secretary for Food Safety, and the Assistant Administrator for Water, respectively, as their outbreak coordinators. However, public health surveillance, both domestic and international, could also be improved through better coordination among public health and law enforcement and *intelligence agencies*. Coordination is inhibited because of the conflicting demands created by the transparency required for public health agencies to operate freely in the United States or abroad, and the requirements of law enforcement or intelligence gathering. Consider, for example, an institution such as a hospital or university that experiences a disease outbreak. Personnel and administrators may talk freely to scientists pursuing a public health mission, but may be much less forthcoming if those investigators are perceived as surrogates for law enforcement agencies which could pursue possible prosecutions. Internationally, the situation is even more delicate. After the plague outbreak in Surat, for example, the Indian newsweekly The Week explicitly accused the United States of being responsible for the outbreak and identified by name four members of the CDC who had arrived in India to study it.

The CDC's desire to send epidemiologists was described as suspicious. U.S. agencies conducting epidemiological or other public health activities, be they civilian or military, will be understandably reluctant to risk compromising their ability to detect and respond to diseases overseas by appearing to have ties with intelligence gathering or covert activities. Nevertheless, the threat of biological terrorism, and potential early ambiguities between natural outbreaks and intentional or accidental releases of biological agents, demand that closer ties between law enforcement, intelligence, and public health be established. For example, public health surveillance could likely benefit from domestic and international intelligence that there was a probable biological threat and consequent concern over the potential use of a particular biological agent. Conversely, law enforcement and intelligence could benefit from being regularly informed on what outbreaks are being seen in public health surveillance (domestic and overseas) and how these events are being resolved.

Federal law enforcement has considerable experience working with local actors (such as local police departments) and success at maintaining the confidentiality of appropriate information passed on in these relationships. The appropriate levels and individuals in the public health surveillance system to receive analogous information need to be determined. Maintenance of confidentiality could be inconsistent with a very broad notification. For this sort of exchange of information to be secure and effective, pre-planning is a requirement.

Role for Scientists and Scientific Societies

Scientists have in the past alerted the public and decision-makers to dilemmas posed by their research, for example in the 1970s during the recombinant DNA controversy. Zilinskas has suggested that, given the difficulties in verifying the BWC, individual scientists and scientific societies must cultivate an ethic in which the illicit development of biological weapons will be discouraged and perhaps revealed to outsiders. He recommends that scientists in nations suspected of sponsoring biological weapons research be especially encouraged to attend scientific meetings and provided with electronic communications access to their international colleagues. Similarly, science students from these nations should be invited to international fora where scientific ethics are discussed. National and international scientific societies such as the American Society for Microbiology and the International Council of Scientific Unions are natural sponsors for such activities. Moreover, scientific and technical workers who once worked in biological weapons programs, for example in the former Soviet Union, should be provided with challenging work in their home nations so as to deter them from marketing their biological weapons skills abroad. International programs established by the European Union, Japan, the United States, and private individuals, such as the International Science and Technology Center, should be supported with these objectives in mind.

National Security and Public Health

Terrorist attacks using *biological weapons* have been carried out or attempted at virtually every scale, from individual assassinations to indiscriminate attacks. While *apocalyptic* urban attacks have not succeeded, they have been attempted by at least one terrorist group. *Prudent national* security policy requires the United States to prepare itself for such attacks.

Because diseases have long incubation times when compared to modern national and international travel times, preparing for *biological terrorism* is necessarily different from preparing for attacks using other weapons of mass destruction. Preparing for bioterrorism requires improving the sensitivity and "*connectivity*" of public health surveillance systems within the United States and overseas. Domestically, physicians and other health care workers must be given the training needed to recognize or at least suspect unusual diseases, and the ability to check these suspicions quickly at the state or regional level must be available. Internationally, the United States should work with foreign governmental, multilateral, and non-governmental organizations to improve global surveillance for suspicious outbreaks. These same systems will also help protect American citizens, and people throughout the world, from emerging diseases.

The *incubation delay* periods of many diseases as well as the growing amount of food imported into the United States demonstrate the insufficiency of protecting the security of U.S. citizens through monitoring for human-borne illnesses at ports of entry or inspection of food imports. Both forms of monitoring are important and should be improved, but some diseases will inevitably elude this screening. There is no alternative to a defense in depth, with improved surveillance at all levels, from the local to the international. The United States has a strong stake in determining the nature and origin of outbreaks overseas. Threatening outbreaks are best recognized and stopped while still abroad.

An important component of such international detection and response will continue to be the dispatching of epidemiological and medical teams overseas. However, international surveillance must primarily rely upon multilateral cooperation, in the form of both sentinel laboratories and formal and informal electronic and voice networks. It is therefore directly in the interest of the United States to strengthen these. Preparations for a biological attack via improved public health surveillance, both domestic and international, will simultaneously protect U.S. citizens against emerging infections and other naturally occurring outbreaks. Even if a major biological terrorist attack never occurs, the investment in public health will, on a daily basis, work to improve the health of all Americans. It is sobering that many of these conclusions are reiterations of lessons learned decades ago.

In 1950, soon after the onset of the Korean War, it was recognized that a biological terrorist attack within the United States was possible, and that little could be done to stop such an attack. However, the disease could be contained and quickly treated if early detection were achieved. To this end, the CDC's *Epidemic Intelligence Service* (EIS) was formed; medical officers were trained in field epidemiology and assigned to the CDC, state health departments, and universities. In the absence of a terrorist attack, members of the EIS could maintain their expertise and improve American public health by analyzing natural disease outbreaks. These baseline assessments would in any case be important for the recognition of possible attacks, as they would be needed to determine whether an outbreak exceeded normal background levels—and was therefore potentially suspicious. An appropriate national security response to the threat of biological terrorism is interwoven with the response that is needed to combat the environmental threat of emerging diseases: improved public health surveillance. In this context, responsibility for national security extends throughout society, from primary care physicians at local hospitals and clinics, to state and national health laboratories and officials, and to public health surveillance networks overseas. These responsibilities need to be matched with appropriate resources. Public health and national security merge in the realm of emerging diseases and biological terrorism.

14

Unbounding the Future

Our bodies are filled with intricate, active molecular structures. When those structures are damaged, health suffers. Modern medicine can affect the workings of the body in many ways, but from a molecular viewpoint it remains crude indeed. *Molecular manufacturing* can construct a range of medical instruments and devices with far greater abilities. The body is an enormously complex world of *molecules*. With *nanotechnology* to help, we can learn to repair it.

Molecular Body

To understand what nanotechnology can do for medicine, we need a picture of the body from a molecular perspective. The human body can be seen as a workyard, construction site, and battleground for *molecular machines*. It works remarkably well, using systems so complex that medical science still doesn't understand many of them. Failures, though, are all too common.

Body as Workyard

Molecular machines do the daily work of the body. When we chew and swallow, muscles drive our motions. Muscle fibers contain bundles of molecular fibers that shorten by sliding past one another.

In the stomach and intestines, the molecular machines we call digestive *enzymes* break down the complex molecules in foods, forming smaller molecules for use as fuel or as building blocks. Molecular devices in the lining of the digestive tract carry useful molecules to the bloodstream.

Meanwhile, in the lungs, molecular storage devices called *hemoglobin* molecules pick up oxygen. Driven by molecular fibers, the

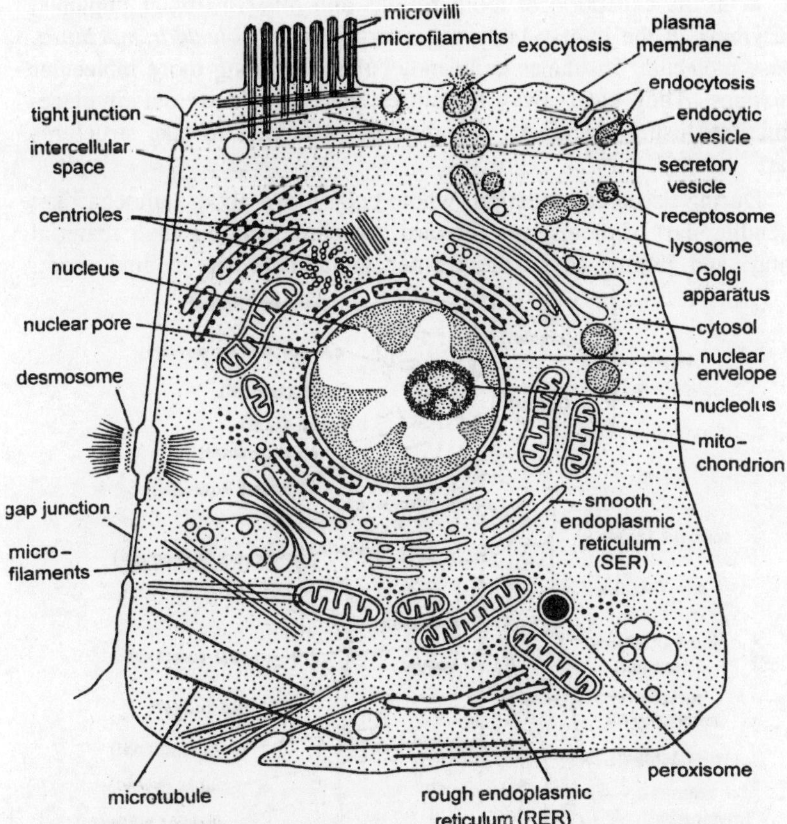

Fig. 14.1. Ultrastructure of a typical animal cells as seen in the electron microscope.

heart pumps blood laden with fuel and oxygen to *cells*. In the muscles, fuel and oxygen drive contraction based on sliding molecular fibers. In the brain, they drive the molecular pumps that charge nerve cells for action. In the liver, they drive molecular macnines that build and break down a whole host of molecules. And so the story continues through all the work of the body. Yet each of these functions sometimes fails, whether through damage or inborn defect.

Body as Construction Site

In growing, healing, and renewing tissue, the body is a construction site. Cells take building materials from the bloodstream. Molecular machinery programmed by the cell's genes uses these materials to build biological structures: to lay down bone and collagen, to build whole new cells, to renew skin, and to heal wounds.

With the exception of tooth fillings and other artificial implants, everything in the human body is constructed by *molecular machines*. These molecular machines build molecules, including more molecular machines. They clear away structures that are old or out of place, sometimes using machinery like digestive enzymes to take structures apart.

During tissue construction, whole cells move about, amoeba like: extending part of themselves forward, attaching, pulling their material along, and letting go of the former attachment site behind them.

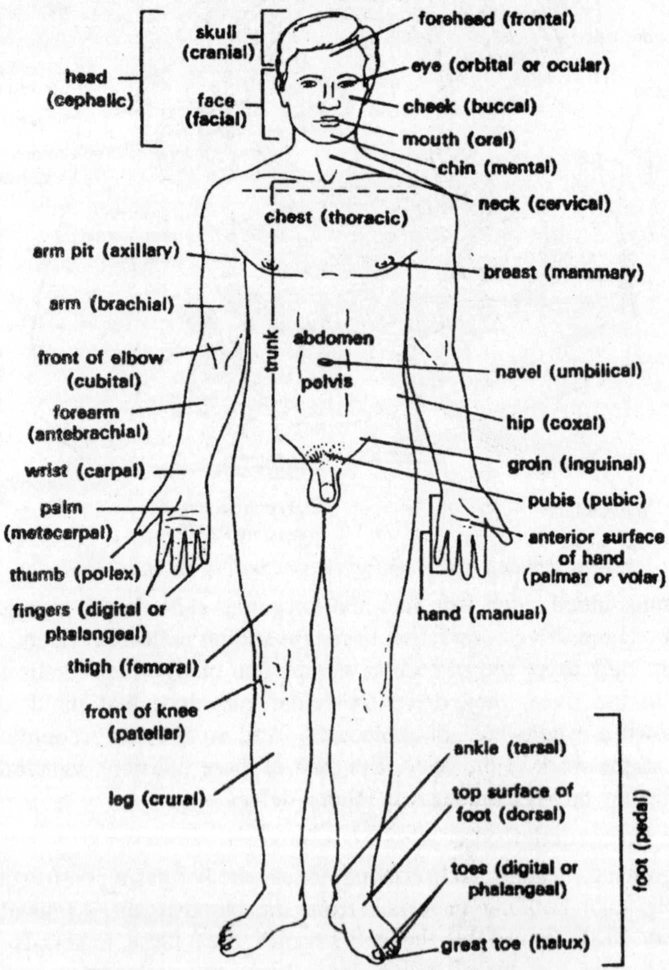

Fig. 14.2. Dorsal view showing the anatomical position.

Individual cells contain a dynamic pattern of molecules made of components that can break down but can also be replaced. Some molecular machines in the cell specialize in digesting molecules that show signs of damage, allowing them to be replaced by fresh molecules made according to genetic instructions. Components inside cells form their complex patterns by self-assembly, that is, by sticking to the proper partners.

Failures in construction increase as we age. Teeth wear and crack and aren't replaced; hair follicles stop working; skin sags and wrinkles. The eye's shape becomes more rigid, ruining close vision. Younger bodies can knit together broken bones quickly, making them stronger than before, but osteoporosis can make older bones so fragile that they break under minor stress.

Sometimes construction is botched from the beginning due to a missing or defective genetic code. In hemophilia, bleeding fails to stop due to the lack of blood clotting factor. Construction of muscle tissue is disrupted in 1 in 3,300 male births by *muscular dystrophy*, in which muscles are gradually replaced by scar tissue and fat; the molecule "*dystrophin*" is missing. *Sickle cell anemia* results from abnormal hemoglobin molecules.

Paraplegics and *quadriplegics* know that some parts of the body don't heal well. The spinal cord is an extreme—and extremely serious—case, but scarring and improper regrowth of tissues result from many accidents. If tissues always regrew properly, injury would do no permanent physical damage.

Body as Battlefield

Assaults from outside the body turn it into a battlefield where the aggressors sometimes get the upper hand. From parasitic worms to protozoa to fungi to *bacteria* to *viruses*, organisms of many kinds have learned to live by entering the body and using their molecular machinery to build more of themselves from the body's building blocks. To meet this onslaught, the body musters the defenses of the immune system—an armada of its own molecular machines. Your body's own amoebalike white blood cells patrol the bloodstream and move out into tissues, threading their way between other cells, searching for invaders.

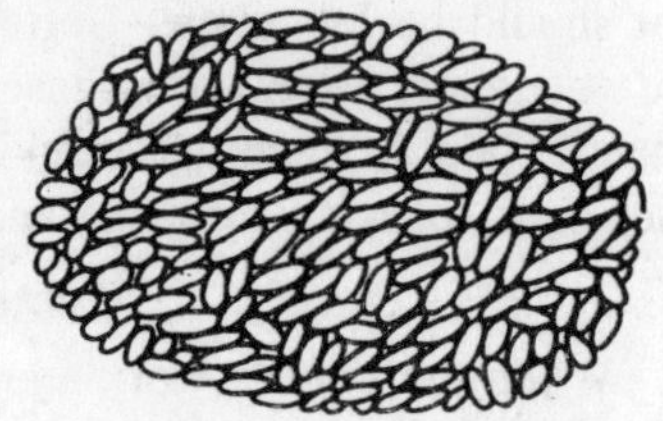

Fig. 14.3. Smallpox virus

How can the immune system distinguish the hundreds of kinds of cells that should be in the body from the invading cells and viruses that shouldn't? This has been the central question of the complex science of immunology. The answer, as yet only partially understood, involves a complex interplay of molecules that recognize other molecules by sticking to them in a selective fashion. These include free-floating antibodies—which are a bit like bumbling guided missiles—and similar molecules that are bound to the surface of white blood cells and other cells of the immune system, enabling them to recognize foreign surfaces on contact.

This system makes life possible, defending our bodies from the fate of meat left at room temperature. Still, it lets us down in two basic ways.

First, the immune system does not respond to all invaders, or responds inadequately. Malaria, tuberculosis, herpes, and AIDS all have their strategies for evading destruction. Cancer is a special case in which the invaders are altered cells of the body itself, sometimes successfully masquerading as healthy cells and escaping detection.

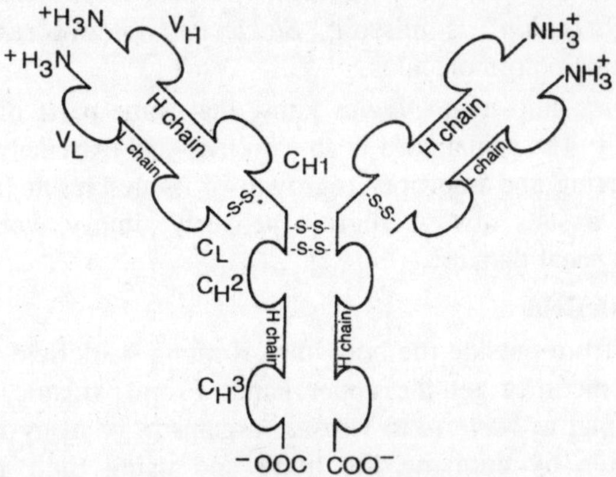

Fig. 14.4. Showing antibody

Second, the immune system sometimes overresponds, attacking cells that should be left alone. Certain kinds of arthritis, as well as lupus and rheumatic fever, are caused by this mistake. Between attacking when it shouldn't and not attacking when it should, the immune system often fails, causing suffering and death.

Medicine Today

When the body's working, building, and battling goes away, we turn to medicine for diagnosis and treatment. Today's methods, though, have obvious shortcomings.

Crude Methods

Diagnostic procedures vary widely, from asking a patient questions, through looking at X-ray shadows, through exploratory surgery and the microscopic and chemical analysis of materials from the body. Doctors can diagnose many ills, but others remain mysteries. Even a diagnosis does not imply understanding: doctors could diagnose infections before they knew about germs, and today can diagnose many syndromes with unknown causes. After years of experimentation and untold loss of life, they can even treat what they don't understand—a drug may help, though no one knows why.

Leaving aside such therapies as heating, massaging, irradiating, and so forth, the two main forms of treatment are surgery and drugs. From a molecular perspective, neither is sophisticated.

Surgery is a direct, manual approach to fixing the body, now practiced by highly trained specialists. Surgeons sew together torn tissues and skin to enable healing, cut out cancer, clear out clogged arteries, and even install pacemakers and replacement organs. It's direct, but it can be dangerous: anesthetics, infections, organ rejection, and missed cancer cells can all cause failure. Surgeons lack fine-scale control. The body works by means of molecular machines, most working inside cells. Surgeons can see neither molecules nor cells, and can repair neither.

Drug therapies affect the body at the molecular level. Some therapies—like insulin for diabetics—provide materials the body lacks. Most—like antibiotics for infections—introduce materials no human body produces. A drug consists of small molecules; in our simulated molecular world, many would fit in the palm of your hand. These molecules are dumped into the body (sometimes directed to a particular region by a needle or the like), where they mix and wander through blood and tissue. They typically bump into other molecules of all sorts in all places, but only stick to and affect molecules of certain kinds.

Antibiotics like penicillin are selective poisons. They stick to molecular machines in bacteria and jam them, thus fighting infection. Viruses are a harder case because they are simpler and have fewer vulnerable molecular machines. Worms, fungi, and protozoa are also difficult, because their molecular machines are more like those found in the human body, and hence harder to jam selectively. Cancer is the most difficult of all. Cancerous growths consist of human cells, and attempts to poison the cancer cells typically poison the rest of the patient as well.

Other drug molecules bind to molecules in the human body and modify their behaviour. Some decrease the secretion of stomach acid, others stimulate the kidneys, many affect the molecular dynamics of the brain. Designing drug molecules to bind to specific targets is a growth industry today, and provides one of the many short-term payoffs that is spurring developments in molecular engineering.

Limited Abilities

Current medicine is limited both by its understanding and by its tools. In many ways, it is still more an art than a science. Mark Pearson of Du Pont points out, "In some areas, medicine has become much more scientific, and in others not much at all. We're still short of what I would consider a reasonable scientific level. Many people don't realize that we just don't know fundamentally how things work. It's like having an automobile, and hoping that by taking things apart, we'll understand something of how they operate. We know there's an engine in the front and we know it's under the hood, we have an idea that it's big and heavy, but we don't really see the rings that allow pistons to slide in the block. We don't even understand that controlled explosions are responsible for providing the energy that drives the machine."

Better tools could provide both better knowledge and better ways to apply that knowledge for healing. Today's surgery can rearrange blood vessels, but is far too coarse to rearrange or repair cells. Today's drug therapies can target some specific molecules, but only some, and only on the basis of type. Doctors today can't affect molecules in one cell while leaving identical molecules in a neighbouring cell untouched because medicine today cannot apply surgical control to the molecular level.

Nanotechnology in Medicine

Developments in *nanotechnology* will result in improved *medical sensors*. As protein chemist Bill DeGrado notes, "Probably the first use you may see would be in diagnostics: being able to take a tiny amount of blood from somebody, just a pinprick, and diagnose for a hundred different things. Biological systems are already able to do that, and we should be able to design molecules or assemblies of molecules that mimic the biological system."

In the longer term, though, the story of nanotechnology in medicine will be the story of extending surgical control to the molecular level. The easiest applications will be aids to the immune system, which selectively attack invaders outside tissues. More difficult applications will require that medical *nanomachines* mimic white blood cells by entering tissues to interact with their cells. Further applications will involve the complexities of molecular-level surgery on individual cells.

As we look at how to solve various problems, you'll notice that some that look difficult today will become easy, while others that might seem easier turn out to be more difficult. The seeming difficulty of treating disorders is always changing: Once polio was frequent and incurable, today it is easily prevented. *Syphilis* once caused steady physical decline leading to insanity and death; now it is cured with a shot.

Athlete's foot has never been seen as a great scourge, yet it remains hard to cure. Likewise with the common cold. This pattern will continue: Deadly diseases may be easily dealt with, while minor ills remain incurable, or vice versa. As we will see, a *mature nanotechnology-based medicine* will be able to deal with almost any physical problem, but the order of difficulty may be surprising. Nature cares nothing for our sense of appropriateness. Horribleness and difficulty just aren't the same thing.

Working Outside Tissues

One approach to *nanomedicine* would make use of microscopic mobile devices built using molecular-manufacturing equipment. These would resemble the *ecosystem protectors* and mobile cleanup machines discussed in the last chapter. Like them, they would either be biodegradable, self-collecting, or collected by something else once they were done working. Like them, they would be more difficult to develop than simple, fixed-location nanomachines, yet clearly feasible and useful. Development will start with the simpler applications, so let's begin by looking at what can be done without entering living tissues.

The *skin* is the body's largest organ, and its exposed position subjects it to a lot of abuse. This exposed position, though, also makes it easier to treat. Among the earlier applications of molecular manufacturing may be those popular, quasimedical products, *cosmetics*. A cream packed with nanomachines could do a better and more selective job of cleaning than any product can today. It could remove the right amount of dead skin, remove excess oils, add missing oils, apply the right amounts of natural moisturizing compounds, and even achieve the elusive goal of "deep pore cleaning" by actually reaching down into pores and cleaning them out. The cream could be a smart material with smooth-on, peel-off convenience.

The mouth, teeth, and gums are amazingly troublesome. Today, daily dental care is an endless cycle of brushing and flossing, of losing ground to tooth decay and gum disease as slowly as possible. A mouthwash full of smart nanomachines could do all that brushing and flossing do and more, and with far less effort—making it more likely to be used.

This mouthwash would identify and destroy pathogenic bacteria while allowing the harmless flora of the mouth to flourish in a healthy ecosystem. Further, the devices would identify particles of food, plaque, or tartar, and lift them from teeth to be rinsed away. Being suspended in liquid and able to swim about, devices would be able to reach surfaces beyond reach of toothbrush bristles or the fibers of floss. As short-lifetime medical nanodevices, they could be built to last only a few minutes in the body before falling apart into materials of the sort found in foods (such as fiber). With this sort of daily dental care from an early age, tooth decay and gum disease would likely never arise. If under way, they would be greatly lessened.

Going beyond this superficial treatment would involve moving among and modifying cells. Let's consider what can be done with this treatment inside the body, but outside the body's tissues. The bloodstream carries everything from nutrients to immune-system cells, with chemical signals and infectious organisms besides.

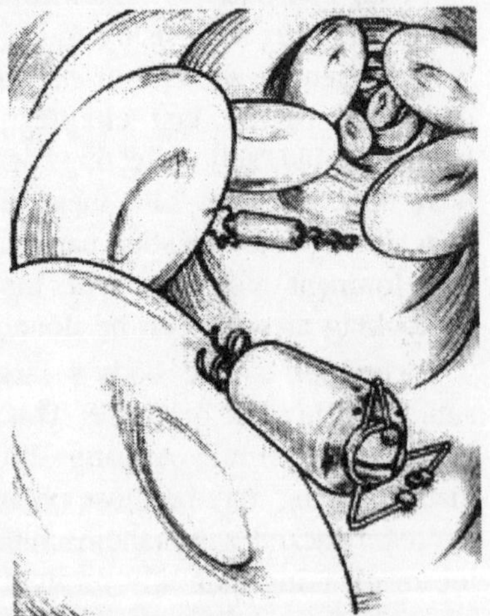

Fig. 14.5. Immune machines.

Here, it is useful to think in terms of medical nanomachines that resemble small submarines. Each of these is large enough to carry a *nanocomputer* as powerful as a mid-1980s mainframe, along with a huge database (a billion bytes), a complete set of instruments for identifying biological surfaces, and tools for clobbering viruses, bacteria, and other invaders. Immune cells, as we've seen, travel through the bloodstream checking surfaces for foreignness and—when working properly—attacking and eliminating what should not be there. These *immune machines* would do both more and less. With their onboard sensors and computers, they will be able to react to the same molecular signals that the immune system does, but with greater discrimination. Before being sent into the body on their search-and-destroy mission, they could be programmed with a set of characteristics that lets them clearly distinguish their targets from everything else. The body's immune system can respond only to invading organisms that had been encountered by that individual's body. Immune machines, however, could be programmed to respond to anything that had been encountered by world medicine.

Immune machines can be designed for use in the bloodstream or the digestive tract (the mouthwash described above used these abilities in hunting down harmful bacteria). They could float and circulate, as antibiotics do, while searching for intruders to neutralize. To escape being engulfed by white blood cells making their own patrols, immune machines could display standard molecules on their surface-molecules the body knows and trusts already—like a fellow police officer wearing a familiar uniform.

When an invader is identified, it can be punctured, letting its contents spill out and ending its effectiveness. If the contents were known to be hazardous by themselves, then the immune machine could hold on to it long enough to dismantle it more completely.

How will these devices know when it's time to depart? If the physician in charge is sure the task will be finished within, say, one day, the devices prescribed could be of a type designed to fall apart after twenty-four hours. If the treatment time needed is variable, the physician could monitor progress and stop action at the appropriate time by sending a specific molecule—aspirin perhaps, or something even safer—as a signal to stop work. The inactivated devices would then be cleared out along with other waste eliminated from the body.

Working Within Tissues

In most parts of the body, the finest blood vessels, capillaries, pass within a few cell diameters of every point. Certain white blood cells can leave these vessels to move among the neighbouring cells. *Immune machines* and similar devices, being even smaller, could do likewise. In some tissues, this will be easy, in some harder, but with careful design and testing, essentially any point of the body should become accessible for healing repairs.

Merely fighting organisms in the bloodstream would be a major advance, cutting their numbers and inhibiting their spread. Roving medical nanomachines, though, will be able to hunt down invaders throughout the body and eliminate them entirely.

Eliminating Invaders

Cancers are a prime example. The immune system recognizes and eliminates most potential cancers, but some get by. Physicians can recognize cancer cells by their appearance and by molecular markers, but they cannot always remove them all through surgery, and often cannot find a selective poison. *Immune machines*, however, will have no difficulty identifying cancer cells, and will ultimately be able to track them down and destroy them wherever they may be growing. Destroying every cancer cell will cure the cancer.

Bacteria, protozoa, worms, and other parasites have even more obvious molecular markers. Once identified, they could be destroyed, ridding the body of the disease they cause. Immune machines thus could deal with tuberculosis, strep throat, leprosy, malaria, amoebic dysentery, sleeping sickness, river blindness, hookworm, flukes, candida, valley fever, antibiotic-resistant bacteria, and even athlete's foot. All are caused by invading cells or larger organisms (such as worms). Health officials estimate that parasitic diseases, common in the Third World, affect more than one billion people. For many of these diseases, no satisfactory drug treatment exists. All can eventually be eliminated as threats to human health by a sufficiently advanced form of nanomedicine.

Herding Cells

Destroying invaders will be helpful, but injuries and structural problems pose other problems. Truly advanced medicine will be able to build up and restructure tissues. Here, medical nanodevices can stimulate and guide the body's own construction and repair mechanisms to restore healthy tissue.

What is healthy tissue? It consists of normal cells in normal patterns in a normal matrix all organized in a normal relationship to the surrounding tissues. Surgeons today (with their huge, crude tools) can fix some problems at the tissue level. A wound disrupts the healthy relationship between two different pieces of tissue, and surgical glues and sutures can partly remedy this problem by holding the tissues in a position that promotes healing. Likewise, coronary artery bypass surgery brings about a more healthy overall configuration of tissues—one that provides working plumbing to supply blood to the heart muscle. Surgeons cut and stitch, but then they must rely on the tissue to heal its wounds as best it can.

Healing establishes healthy relationships on a finer scale. Cells must divide, grow, migrate, and fill gaps. They must reorganize to form properly connected networks of fine blood vessels. And cells must lay down materials to form the structural, intercellular matrix—collagen to provide the proper shape and toughness, or mineral grains to provide rigidity, as in bone. Often, they lay down unwanted scar tissue instead, blocking proper healing.

With enough knowledge of how these processes work (and *nanoinstruments* can help gather that knowledge) and with good enough software to guide the process—a more difficult challenge—medical nanomachines will be able to guide this healing process. The problem here is to guide the motion and behaviour of a mob of active, living cells—a process that can be termed *cell herding*.

Cells respond to a host of signals from their environment: to chemicals in the surrounding fluids, to signal molecules on neighbouring cells, and to mechanical forces applied to them. Cell-herding devices would use these signals to spur cell division where it is needed and to discourage it where it is not. They would nudge cells to encourage them to migrate in appropriate directions, or would simply pick them up, move them along, and deliver them where needed, encouraging them to nestle into a proper relationship with their neighbours. Finally, they would stimulate cells to surround themselves with the proper intercellular-matrix materials. Or—like the owner of a small dog who, on a cold day, wraps the beast in a wool jacket—they would directly build the proper surrounding structures for the cell in its new location.

In this way, cooperating teams of cell-herding devices could guide the healing or restructuring of tissues, ensuring that their cells form healthy patterns and a healthy matrix and that those tissues have a healthy relationship to their surroundings. Where necessary, cells could even be adjusted internally, as we will discuss later.

Rebuilding Tissues

Again, skin provides easy examples and may be a natural place to start in practice. People often want hair where they have bare skin, and bare skin where they have hair. *Cell herding machines* could move or destroy hair follicle cells to eliminate an unwanted hair, or grow more of the needed cells and arrange them into a working follicle where a hair is desired. By adjusting the size of the follicle and the properties of some of the cells, hairs could be made coarser, or finer, or straighter, or curlier. All these changes would involve no pain, toxic chemicals, or stench. Cell-herding devices could move down into the living layers of skin, removing unwanted cells, stimulating the growth of new cells, narrowing unnaturally prominent blood vessels,

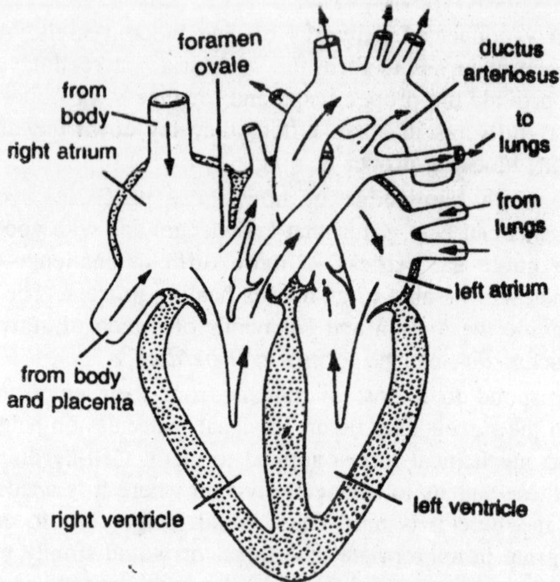

Fig. 14.6. Heart showing direction of blood flow.

insuring good circulation by guiding the growth of any needed normal blood vessels, and moving cells and fibers around so as to eliminate even deep wrinkles.

At the opposite end of the spectrum, cell herding will revolutionize treatment of life-threatening conditions. For example, the most common cause of heart disease is reduced or interrupted supply of blood to the heart muscle. In pumping oxygenated blood to the rest of the body, the heart diverts a portion for its own use though the coronary arteries. When these blood vessels become constricted, we speak of coronary-artery disease. When they are blocked, causing heart muscle tissue to die, we speak of someone "having a coronary," another term for heart attack.

Devices working in the bloodstream could nibble away at atherosclerotic deposits, widening the affected blood vessels. Cell herding devices could restore artery walls and artery linings to health, by ensuring that the right cells and supporting structures are in the right places. This would prevent most heart attacks.

But what if a heart attack has already destroyed muscle tissue, leaving the patient with a scarred, damaged, and poorly functioning heart? Once again, cell-herding devices could accomplish repairs, working their way into the scar tissue and removing it bit by bit, replacing it with fresh muscle fiber. If need be, this new fiber can be grown by applying a series of internal molecular stimuli to selected heart muscle cells to "remind" them of the instructions for growth that they used decades earlier during embryonic development.

Cell-herding capabilities should also be able to deal with the various forms of arthritis. Where this is due to attacks from the body's own immune system, the cells producing the damaging antibodies can be identified and eliminated. Then a cell-herding system would work inside the joint where it would remove diseased tissues, calcified spurs, and so forth, then rework patterns of cells and intercellular material to form a healthy, smoothly working, and pain-free joint. Clearly, learning to repair hearts and learning to repair joints will have some basic technologies in common, but much of the research and development will have to be devoted to specific tissues and specific circumstances. A similar process—but again, specially adapted to the circumstances at hand—could be used to strengthen and reshape bone, correcting osteoporosis.

In *dentistry*, this sort of process could be used to fill cavities, not with amalgam, but with natural dentin and enamel. Reversing the ravages of periodontal disease will someday be straightforward, with nanomedical devices to clean pockets, join tissues, and guide regrowth. Even missing teeth could be regrown, with enough control over cell behaviour.

Working on Cells

Moving through tissues without leaving a trail of disruption will require devices able to manipulate and direct the motions of cells, and to repair them. Much remains to be learned—and will be easy to learn with *nanoscale tools*—but today's knowledge of cells is enough for a start on the problem of how to do surgery on cells.

Cell biology is a booming field, even today. Cells can be made to live and grow in laboratory cultures if they are placed in a liquid with suitable nutrients, oxygen, and the rest. Even with today's crude techniques, much has been learned about how cells respond to different chemicals, to different neighbours, and even to being poked and cut with needles. Conducting a rough sort of surgery on individual cells has been routine for many years in scientific laboratories.

Today, researchers can inject new *DNA* into cells using a tiny needle; small punctures in a cell membrane automatically reseal. But both these techniques use tools that on a cellular scale are large and clumsy—like doing surgery with an ax or a wrecking ball, instead of a scalpel. *Nanoscale* tools will enable medical procedures involving delicate surgery on individual cells.

Eliminating Viruses by Cell Surgery

Some viral diseases will respond to treatments that destroy viruses in the nose and throat, or in the bloodstream. The flu and common cold are examples. Many others would be greatly improved by this, but not eliminated. All viruses work by injecting their genes into a cell and taking over its molecular machinery, using it to produce more viruses. This is part of what makes viral illnesses so hard to treat—most of the action is performed by the body's own molecular machines, which can't be interfered with on a wholesale basis. When the immune system deals with a viral illness, it both attacks free virus particles before they enter cells, and attacks infected cells before they can churn out too many more virus particles.

Some viruses, though, insert their genes among the genes of the cell, and lay low. The cell can seem entirely normal to the immune system, for months or years, until the viral genes are triggered into action and begin the infective process anew. This pattern is responsible for the persistence of herpes infections, and for the slow, deadly progress of AIDS. These viruses can be eliminated by molecular-level cellular surgery. The required devices could be small enough to fit entirely *within* the cell, if need be. Greg Fahy, who heads the *Organ Cryopreservation Project* at the American Red Cross's Jerome Holland Transplantation Laboratory, writes, "Calculations imply that molecular sensors, molecular computers, and molecular effectors can be combined into a device small enough to fit easily inside a single cell and powerful enough to repair molecular and structural defects (or to degrade foreign structures such as viruses and bacteria) as rapidly as they accumulate.... There is no reason such systems cannot be built and function as designed."

Equally well, a *cell surgery* device located outside a cell could reach through the membrane with long probes. At the ends of the probes would be tools and sensors along with, perhaps, a small auxiliary computer. These would be able to reach through multiple membranes, unpackage and uncoil DNA, read it, repackage it, and recoil it, "proofreading" the DNA by comparing the sequences in one cell to the sequences of other cells.

On reading the genetic sequence spelling out the message of the AIDS virus, a *molecular surgery* machine could be programmed to respond like an immune machine, destroying the cell. But it would seem to make more sense simply to cut out the AIDS virus genes themselves, and reconnect the ends as they were before infection. By doing this, and killing any viruses found in the cell, the procedure would restore the cell to health.

Molecular Repairs

Cells are made of billions of molecules, each built by *molecular machines*. These molecules self-assemble to form larger structures, many in dynamic patterns, perpetually disintegrating and reforming. Cell-surgery devices will be able to make molecules of sorts that may be lacking, while destroying molecules that are damaged or present in excess. They will be able not only to remove viral genes, but to repair chemical and radiation-caused damage to the cell's own genes. Advanced cell surgery devices would be able to repair cells almost regardless of their initial state of damage.

By activating and inactivating a cell's genes, they will be able to stimulate cell division and guide what types of cells are formed. This will be a great aid to cell herding and to healing tissues.

As surgeons today rely on the spontaneous, self-organizing ability of cells and tissues to join and heal the parts they manipulate, so cell-surgery devices will rely on the spontaneous self-organizing capabilities of molecules to join and "heal" the parts they put together. Healing of a surgical wound involves sweeping up dead cells, growing new cells, and a slow and genuinely painful process of tissue reorganization. In contrast, the joining of molecules is almost instantaneous and occurs on a scale far below that of the most sensitive pain receptor. "Healing" will not begin *after* the repair devices have done their work, as it does in conventional surgery: rather, when they complete their work, the tissue will have been healed.

Healing Body and Limb

The ability to herd cells and to perform *molecular repairs* and cell surgery will open new vistas for medicine. These abilities apply on a small scale, but their effects can be large scale.

Correcting Chemistry

In many diseases, the body as a whole suffers from misregulation of the signaling molecules that travel through its fluids. Many are rare: Cushing's disease, Grave's disease, Paget's disease, Addison's disease, Conn's syndrome, Prader-Labhart-Willi syndrome. Others are common: millions of older women suffer from osteoporosis, the weakening of bones that can accompany lowered estrogen levels.

Diabetes kills frequently enough to rank in the top ten causes of death in the United States; the number of individuals known to have it doubles every fifteen years. It is the leading cause of blindness in the United States, with other complications including kidney damage, cataracts, and cardiovascular damage. Today's *molecular medicine* tries to solve these troubles by supplying missing molecules: diabetics inject additional insulin. While helpful, this doesn't cure the disease or eliminate all symptoms. In an era of molecular surgery, physicians could choose instead to repair the defective organ, so it can regulate its own chemicals again, and to readjust the metabolic properties of other cells in the body to match. This would be a true healing, far better than today's partial fix.

Only now are researchers making progress on another frequent problem of metabolic regulation: obesity. Once this was thought to have one simple cause (consuming excess calories) and one main result (greater roundness than favoured by today's aesthetics), but both assumptions proved wrong. Obesity is a serious medical problem, increasing the risk of diabetes mellitus, osteoarthritis, degenerative diseases of the heart, arteries, and kidneys, and shortening life expectancy. And the supposed cause, simple overeating, has been shown to be incorrect—something dieters had always suspected, as they watched thinner colleagues gorge and yet gain no weight.

The ability to lay in stores of fat was a great benefit to people once upon a time, when food supplies were irregular, nomadism and marauding bands made food storage difficult and risky, and starvation was a common cause of death. Our bodies are still adapted to that world, and regulate fat reserves accordingly. This is why dieting often has perverse effects. The body, when starved, responds by attempting to build up greater reserves of fat at its next opportunity. The main effect of exercise in weight reduction isn't to burn up calories, but to signal the body to adapt itself for efficient mobility.

Obesity therefore seems to be a matter of chemical signals within the body, signals to store fat for famine or to become lean for motion. Nanomedicine will be able to regulate these signals in the bloodstream, and to adjust how individual cells respond to them in the body. The latter would even make possible the elusive "spot reduction program" to reshape the distribution of body fat.

Here, as with many potential applications of nanotechnology, the problem may be solved by other means first. Some problems, though, will almost surely require nanomedicine.

New Organs and Limbs

So far we've seen how *medical nanotechnology* would be used in the simpler applications outside tissues—such as in the blood—then inside tissues, and finally inside cells. Consider how these abilities will fit together for victims of automobile and motorcycle accidents.

Nanomanufactured medical devices will be of dramatic value to those who have suffered massive trauma. Take the case of a patient with a crushed or severed spinal cord high in the back or in the neck. The latest research gives hope that when such patients are treated promptly after the injury, paralysis may be at least partially avoidable, sometimes. But those whose injuries weren't treated—including virtually all of today's patients—remain paralyzed. While research continues on a variety of techniques for attempting to aid a spontaneous healing process, prospects for reversing this sort of damage using conventional medicine remain bleak.

With the techniques discussed above, it will become possible to remove scar tissue and to guide cell growth so as to produce healthy arrangements of the cells on a microscopic scale. With the right molecular-scale poking and prodding of the cell nucleus, even nerve cells of the sorts found in the brain and spinal cord can be induced to divide. Where nerve cells have been destroyed, there need be no shortage of replacements. These technologies will eventually enable medicine to heal damaged spinal cords, reversing paralysis.

The ability to guide cell growth and division and to direct the organization of tissues will be sufficient to regrow entire organs and limbs, not merely to repair what has been damaged. This will enable medicine to restore physical health despite the most grievous injuries.

If this seems hard to believe, recall that medical advances have shocked the world before now. To those in the past, the idea of cutting people open with knives *painlessly* would have seemed miraculous, but surgical anesthesia is now routine. Likewise with bacterial infections and antibiotics, with the eradication of smallpox, and the vaccine for polio: Each tamed a deadly terror, and each is now half-forgotten history. Our gut sense of what seems likely has little to do with what can and cannot be done by medical technology. It has more to do with our habitual fears, including the fear of vain hopes. Yet what amazes one generation seems obvious and even boring to the next. The first baby born after each breakthrough grows up wondering what all the excitement was about. Besides, nanoscale medicine won't be a cure-all. Consider a fifty-year-old mentally retarded man, with a mind like a two-year-old's, or a woman with a brain tumour that has spread to the point that her personality has changed: How could they be "healed"? No healing of tissues could replace a missed lifetime of adult experience, nor can it replace lost information from a severely damaged brain. The best physicians could do would be to bring the patients to *some* physically healthy condition. One can wish for more, but sometimes it won't be possible.

First Aid

Throughout the centuries, medicine has been constrained to maintain functioning tissues, since once tissues stop functioning, they can't heal themselves. With molecular surgery to carry out the healing directly, medical priorities change drastically—function is no longer absolutely necessary. In fact, a physician able to use molecular surgery would prefer to operate on nonfunctioning, structurally stable tissue than on tissue that has been allowed to continue malfunctioning until its structure was lost.

Brain tumours are an example: They destroy the brain's structure, and with it the patient's skills, memories, and personality. Physicians in the future should be able to immediately interrupt this process, to stop the functioning of the brain to stabilize the patient for treatment.

Techniques available today can stop tissue function while preserving tissue structure. Greg Fahy, in his work on organ preservation at the American Red Cross, is developing a technique for vitrifying animal kidneys—making them into a low-temperature, crystal-free glass—with the goal of maintaining their structure such that, when brought back to room temperature, they can be transplanted. Some kidneys have been cooled to -30°C, warmed back up, and then functioned after transplantation.

A variety of other procedures can also stabilize tissues on a long-term basis. These procedures enable many cells—but not whole tissues—to survive and recover without help; advanced molecular repair and cell surgery will presumably tip the balance, enabling cells, tissues, and organs to recover and heal. When applied to stabilizing a whole patient, such a condition can be called *biostasis*. A patient in biostasis can be kept there indefinitely until the required medical help arrives. So in the future, the question "Can this patient be restored to health?" will be answered "Yes, if the patient's brain is intact, and with it the patient's mind."

Sandra Lee Adamson of the National Space Society has her eyes on distant goals. Some have proposed that travel to the stars would take generations, preventing anyone on Earth from ever making the trip. But she notes that biostasis will "give hope to some fearless adventurers who will risk suspension and subsequent reanimation so they can see the stars for themselves."

Plague Insurance

Medical nanotechnologies promise to extend healthy life, but if history is any guide, they may also avert sudden massive death. The word *plague* is rarely heard today, except in relation to AIDS; it calls up visions of the Black Death of the Middle Ages, when one third of Europe died in 1346-50. A virulent influenza struck in 1918, half lost in the news of the First World War: how many of us realize that it killed at least 20 million? People often act as though plagues were gone for good, as if sanitation and antibiotics had vanquished them. But as doctors are forever telling their patients, antibiotics kill bacteria, but are useless for viruses. The flu, the common cold, herpes, and AIDS—none has a really effective treatment, because all are caused by viruses. In some African countries, as much as 10 percent of the population is estimated to be infected with the AIDS-causing HIV virus. Without a cure soon, the steep rise in deaths from AIDS still lies in the future. AIDS stands as a grim reminder that the great plagues of history are not behind us.

Threat

New diseases continue to appear today as they have throughout history. Today's population, far larger than that of any previous century, provides a huge, fertile territory for their spread.

Today's transportation systems can spread viruses from continent to continent in a single day. When ships sailed or churned their way across the seas, an infected passenger was likely to show full-blown disease before arrival, permitting quarantine. But few diseases can be guaranteed to show themselves in the hours of a single aircraft flight.

So far as is known, every species of organism, from bacterium to whale, is afflicted with viruses. Animal viruses sometimes "jump the species gap" to infect other animals, or people. Most scientists believe that the ancestors of the AIDS virus could, until recently, infect only certain African monkeys. Then these viruses made the interspecies jump. A similar jump occurred in the 1960s when scientists in West Germany, working with cells from monkeys in Uganda, suddenly fell ill. Dozens were infected, and several died of a disease that caused both blood clots and bleeding, caused by what is now named the

Marburg virus. What if the Marburg virus had spread with a sneeze, like influenza or the common cold?

We think of human plagues as a health problem, but when they hit our fellow species, we tend to see them from an environmental perspective. In the late 1980s, over half the harbor-seal population in large parts of the North Sea suddenly died, leading many at first to blame pollution. The cause, though, appears to be a distemper virus that made the jump from dogs. Biologists worry that the virus could infect seal species around the world, since distemper virus can spread by aerosols—that is, by coughing—and seals live in close physical contact. So far its mortality rate has been 60 to 70 percent.

What of AIDS itself: Could it change and give rise to a form able to spread, say, as colds do? Nobel Laureate Howard M. Temin has said, "I think that we can very confidently say that this can't happen." Nobel Laureate Joshua Lederberg, president of Rockefeller University in New York City, replied, "I don't share your confidence about what can and cannot happen." He points out that "there is no reason a great plague could not happen again. We live in evolutionary competition with microbes—bacteria and viruses. There is no guarantee that we will be the survivors."

Inadequate Abilities

Bacterial diseases are mostly controllable today. Sanitation limits the ways in which plague can spread. These measures are just good enough to lull us into imagining the problem is solved.

Viruses are common, viruses mutate; some spread through the air, and some are deadly. Plagues show that fast-spreading diseases can be deadly, and effective antiviral drugs are still rare.

The only really effective treatments for viral diseases are preventive, not curative. They work either by preventing exposure, or by exposing the body beforehand to dead or harmless or fragmentary forms of the virus, to prepare the immune system for future exposure. As the long struggle for an AIDS vaccine shows, one cannot count on modern medicine to identify a new virus and produce an effective vaccine within a single month or year or even a single decade. But influenza epidemics spread fast, and Marburg II or AIDS II or something entirely new and deadly may do the same.

Doing Better

The deaths from the next great plague could have begun in a village last week, or could begin next year, or a year before we learn to deal with new viral illnesses promptly and effectively. With luck, the plague will wait until a year after.

Immune machines could be set to kill a new virus as soon as it is identified. The instruments nanotechnology brings will make viral identification easy. Some day, the means will be in place to defend human life against viral catastrophe. From eliminating viruses to repairing individual cells, improving our control of the molecular world will improve health care. Immune machines working in the bloodstream seem about as complex as some engineering projects human beings have already completed—projects like large satellites, Other medical nanotechnologies seem to be of a higher order of complexity.

Solving Hard Problems

Somewhere in the progression from relatively simple immune devices to molecular surgery, we've crossed the fuzzy line between systems that teams of clever biomedical engineers could design in a reasonable length of time and ones that might take decades or prove impossibly complex. Designing a nanomachine capable of entering a cell, reading its DNA, finding and removing a deadly viral DNA sequence, and then restoring the cell to normal would be a monumental job. Such tasks are advanced applications of nanotechnology, far beyond mere computers, manufacturing equipment, and half-witted "*smart materials*."

To succeed within a reasonable number of years, we may need to automate much of the engineering process, including software engineering. Today's best expert systems are nowhere near sophisticated enough. The software must be able to apply physical principles, engineering rules, and fast computation to generate and test new designs. Call it *automated engineering*.

Automated engineering will prove useful in advanced nanomedicine because of the sheer number of small problems to be solved. The human body contains hundreds of kinds of cells forming a huge number of tissues and organs. Taken as a whole (and ignoring the immune system), the body contains hundreds of thousands of different kinds of molecules. Performing complex molecular repairs on a damaged cell might require solving millions of separate, repetitive problems. The molecular machinery in cell surgery devices will need to be controlled by complex software, and it would be best to be able to delegate the task of writing that software to an automated system. Until then, or until a lot of more conventional design work gets done, nanomedicine will have to focus on simpler problems.

Aging

Where does aging fit in the spectrum of difficulty? The deterioration that comes with aging is increasingly recognized as a form of disease, one that weakens the body and makes it susceptible to a host of other diseases. Aging, in this view, is as natural as smallpox and bubonic plague, and more surely fatal. Unlike bubonic plague, however, aging results from internal malfunctions in the molecular machinery of the body, and a medical condition with so many different symptoms could be complex.

Surprisingly, substantial progress is being made with present techniques, without even a rudimentary ability to perform cell surgery in a medical context. Some researchers believe that aging is primarily the result of a fairly small number of regulatory processes, and many of these have already been shown to be alterable. If so, aging may be tackled successfully before even simple cell repair is available. But the human aging process is not well enough understood to enable a confident projection of this; for example, the number of regulatory processes is not yet known. A thorough solution may well require advanced nanotechnology-based medicine, but a thorough solution seems possible. The result would not be immortality, just much longer, healthier lives for those who want them.

Restoring Species

A challenging problem related to medicine (and to biostasis) is that of species restoration. Today, researchers are carefully preserving samples from species now becoming extinct. In some cases, all they have are tissue samples. For other species, they've been able to save germ cells in the hope that they will be able to implant fertilized eggs into related species and thus bring the (nearly?) extinct species back.

Each cell typically contains the organism's complete genetic information, but what can be done with this? Many researchers today collect samples for preservation thinking only of the implantation scenario: one that they know has already been made to work. Other researchers are taking a broader view: the Center for Genetic Resources and Heritage at the University of Queensland is a leader in the effort. Daryl Edmondson, coordinator of the gene library, explains that the center is unique because it will "actively collect data. Most other libraries simply collate their own collections." Director John Mattick describes it as a "genetic Louvre" and points out that if genes from today's endangered species aren't preserved, "subsequent generations will see we had the technology to keep [DNA] software and will ask why we didn't do it." With this information and the sorts of molecular repair and cell-surgery capabilities we have discussed, lost species can someday be returned to active life again as habitats are restored.

One such center isn't enough: the Queensland center focuses on Australian species (naturally enough) and has limited funds. Besides, anything so precious as the genetic information of an endangered species should be stored in many separate locations for safety. We need to take out an insurance policy on Earth's genetic diversity with a broader network of genetic libraries, concentrating special attention on gathering biological samples from the fast-disappearing rain forests. Scientific study can wait: the urgency of the situation calls for a vacuum-cleaner approach. The Foresight Institute is promoting this effort through its BioArchive Project.

15

COSTS AND CONSEQUENCES

Until quite recently the massive U.S. post-9/11 biodefense buildup has been discussed almost exclusively in terms of its presumed necessity, the substantial improvements achieved in response readiness, the successive increase of expenditure and broadening programs, and urgings from various quarters for still further increases in expenditure and expansion of programs. Certainly, some of this was needed, and others have analyzed the efficacy or performance of various elements of U.S. biodefense expansion since 2001. However, these programs should be justified on their intrinsic merits, and not due to alleged "*spin-off*" benefits for generic public health. "*Spin-off*" rationalizations for defense R&D expenditure historically have been spurious. They were made for the U.S. BW program over 35 years ago as well.

The suggested "*spin off*" can always be procured for a fraction of the cost of whatever the larger parent program may be, by direct, targeted investments. There are, however, also costs. The first of these is direct federal expenditure. Dr. David Franz is fond of pointing out that in 1996 and 1997, after the three major BW disclosures—of the USSR's enormous covert and illegal BW program, the Iraqi BW program, and the failed attempts of the Japanese Aum Shinrikyo group to produce BW agents—U.S. biodefense expenditures were still in the range of roughly $150 million per year. They increased to $414 million by FY 2001. It was estimated at $7.5 billion for 2005. Annual *civilian* biodefense expenditure has risen more than 18-fold and has accounted for over $22 billion in expenditure during the past 4 fiscal years. The U.S. FY 2006 civilian *biodefense* budget adds another $4.2 billion. The question, of course, is whether this degree of expenditure is merited. That is where the threat assessment should be the crucial determinant. Cumulative DoD biodefense expenditure for the past 4 years is not available. The budget for the joint DoD CB defense program was $1.25 billion in FY 2004, $1.38 billion in FY 2005, and $1.6 billion in FY 2006.

The paradox of this situation is that this change in U.S. Government priorities is primarily due to the events of 9/11, which had no relation whatsoever to the capability to produce biological agents by terrorist groups. And as has already been indicated, the significance of the U.S. *anthrax* events in regard to the anticipation of future events of the same nature carried out by terrorist groups is also unclear. In carrying out the 9/11 aircraft attacks in the United States, the al-Qaida organization certainly was able to demonstrate its enterprise, ingenuity and organizational capabilities—as well as, we now know, a modicum of luck and the failure of various existing U.S. Governmental functions. At the same time, it demonstrated that the group had not been spending the major portion of its time and effort to develop *biological weapons*. As a preceding section of this monograph indicates, as best is known, little regarding *al-Qaida* and BW has changed since.

It is very possible that the U.S. political response and the congressional funding levels and programs that followed, would have been substantially smaller if it were known for certain that the Amerithrax anthrax had been prepared by a U.S. professional, or had been diverted from stocks prepared within the U.S. biodefense program. On the other hand, it undoubtedly would also have led to greater oversight of the U.S. *biodefense program*. Another predictable cost has been the impact on other U.S. public-health programs and expenditures. Currently one-third of both the National Institutes of Health (NIH) infectious disease budget and the Centers for Disease Control and Prevention (CDC) budget and more than half of U.S. Government and corporate vaccine development is relegated to biodefense, that is, it focuses on the "*select agents*," those pathogens that are considered most likely to be used as biological weapon agents.

Equally or more striking were tallies produced in February and March 2005 of changes in the funding patterns of the *National Institutes for Allergy and Infectious Diseases* (NIAID) at NIH in the years between 1996 and 2000 compared to 2001 to 2004. Grants for research on six bacterial pathogens on the select agent list grew from 33 in the first 4r-year period to 497 in the second. Tabulations for research on viral pathogens were similar, except for research on *influenza*. Grants for all other agents dropped between 20 and 50 percent just between FY

1999 to 2001 and FY 2002 to 2004, including for tuberculosis and *acquired immune deficiency syndrome* (AIDS). There has been a 30-fold increase in NIH-NIAID biodefense expenditure since 2001. It now accounts for over 35 percent of the current NIH-NIAID budget, an amount greater than for AIDS research and greater than for all other non-AIDS infectious disease research.

An extremely sharp attack on this shift in public health priorities appeared in the *American Journal of Public Health* in October 2004. At the end of May 2005, NIH announced a shift in disbursement schedules for grants that would also lead to an earlier termination than planned for grants for some research projects on malaria, HIV/AIDS and other *infectious diseases*.

In a May 2005 report on U.S. preparations for a pandemic flu outbreak, the U.S. GAO pointed out that "the Department of Health and Human Services has not finalized planning for an influenza pandemic. In 2000, GAO recommended that DHHS complete the national plan for responding to an influenza pandemic, but the plan has been in draft format since August 2004." At the time of the 2000 report, GAO also took the DHHS to task for lack of progress in developing a vaccine against H5N1 flu. At his retirement on December 3, 2004, DHHS Secretary Tommy Thompson cited pandemic flu as the greatest threat to be faced; yet the situation is virtually the same today as it was in 2000. Washington policymakers have had almost 9 years since the first outbreak of avian flu in 1997 to come to grips with the problem Instead, the focus has been on "*bioterrorism*" and *biodefense*. The U.S. CDC has offered one estimate of the consequences of a "*medium level*" flu pandemic outbreak in the United States "in the absence of any control measures" (e.g., *vaccination* and drugs):

1. 15 to 35 percent of the U.S. population infected;
2. 20 to 47 million cases of illness;
3. 18 to 42 million outpatient hospital visits;
4. 89,000 to 207,000 deaths; and,
5. "associated costs ranging from $71 billion to $167 billion."

The estimate of mortality appears to be low, given that its lower level is about the same as the upper level of ordinary annual U.S. *flu* mortality. Nevertheless, even weeks after the above information was presented to Congress, U.S. senators and congressmen, both Democrat and Republican, were still focused on drumming up support for further increases in Federal expenditures against "*Bioterrorism*" to support "*Bioshield II*," subsidies for medical countermeasures against "*select agents*." Only in November 2005 did the administration finally announce a plan and accompanying recommendations for expenditures to prepare for and to combat pandemic flu.

Biodefense Research and the Biological Weapons Convention

The third area of cost concerns *biological weapons* arms control. This was treated to some degree in *The Problem of Biological Weapons*, and summarized in a subsequent paper by Jonathan Tucker. Tucker makes two major points. The first concerns proliferation, and it is that "The most serious risk associated with science-based threat assessment is that the novel pathogens and information it generates could leak out to rogue states and terrorists." The risk may be less a "*leak*" in the classic sense than simply the accelerated accretion of relevant science and publications, and the substantial overall push that the field is now getting and which will continue in the coming years. Both the Aum Shinrikyo and the al-Qaida groups went back to look at professional literature of previous decades. It is the same procedure that new or expanding state programs followed, whether it was Russia in the 1950s and 1960s, or Iraq in the 1980s.

Tucker's second main point is that the greatly increased magnitude of the U.S. biodefense R&D program will promote a BW arms race, and, at least on the part of others, perhaps not all of it of a defensive nature. The same point was made by Dr. Malcolm Dando in a submission to the British Parliament in February 2003. That arms race, at least in its initial stages, is more likely to be with developments in our own BW research program than against developments in the programs of other states or nonstate actors. This is exactly the process that took place in the United States from the late 1950s to the mid-1970s regarding development of intercontinental ballistic missile (ICBM) re-entry vehicles (warheads) and ABM systems. It was succinctly described by Dr. Jerome B. Wiesner, President John F. Kennedy's "*Science Adviser*", who had been involved with policy planning for these antagonistic weapon systems for years. It was a process in which development in our own offensive and defensive strategic nuclear missile systems fed off the certain knowledge of developments in the other. The "*intelligence*" was much more certain than guessing about what was going on in other nations' analogous programs, and could always be assumed, or attributed, to them. At times, such attribution was correct, and at times not, but even in the former case, new technological developments in the U.S. programs quite frequently were made well in advance of when they appeared elsewhere. Overall, the outcome was the same as that posited by Tucker: the stimulation of parallel programs in other states. The same very likely will occur now. At the present time, one can assume that a smaller replica of the U.S. biodefense program is taking place in Russia, smaller because of the great disparity in funding levels. Russia retains the personnel and facilities to build on their own work dating from their accelerated 1973 to 1992 (or longer) offensive program, as well as the ability to pick up from developments in the United States and in the nonmilitary published literature on the functioning of the human immune system, etc. In some cases, the United States is currently funding research in Russia that is BW applicable.

The fourth cost also concerns arms control but is sufficiently significant and different to require separate consideration. It is the question of whether the U.S. Government, because of the biodefense R&D program, remains in compliance with the provisions of the Biological Weapons Convention (BWC). Towards the end of 1999, U.S. Director of Central Intelligence George Tenet had established a Non-Proliferation Advisory Group (NAG) to advise the CIA on what kinds of research it should undertake in order to better understand the problems that the agency faced in learning about WMD proliferation and, if possible, hindering it. In one of its meetings, NAG was given a briefing on a particular CIA BW-related project, code-named "*clear vision*" It involved the fabrication and testing of a model of a Soviet BW bomblet. It appears that the project was already underway since 1997. One of the members of NAG was Dr. Joshua Lederberg, who has served as an adviser and consultant for BW issues to U.S. Government agencies for the past 4 decades. After hearing the briefing, Dr. Lederberg raised two considerations:

1. that the project raised BWC compliance issues,
2. that the project raised perceptual issues; that is, if the project subsequently became publicly known, it would raise questions in the view of observers as to whether the U.S. Government was engaged in activities of an offensive BW character.

He therefore suggested that the CIA Director could not authorize such a project on his own authority; it would have to be referred to the office of the President for consideration, and to undergo interagency review. Another member of NAG offered a third consideration: that if U.S. intelligence agencies discovered that another country was carrying out such a project, it would be considered prima facie evidence of the existence of an offensive BW program in that country. The project was then referred to the *National Security Council* (NSC) for review. It was nevertheless ultimately approved, over the minority objections of the legal adviser in the U.S. Department of State. It would be carried out at a classified level.

As indicated, it appears the project was already in process by the time the briefing was given, and before the NSC review took place, and that more than one bomblet was actually produced in order to carry out different tests at different sites. None of four such projects—*Clear vision*, *Bacus*, *Bite size*, and *Jefferson*—were reported by the United States in its annual Confidence Building Measures submissions under the terms of the BWC. The ongoing utilization of several very large aerosol test chambers in U.S. biodefense projects was also not reported. In the spring of 2004, a briefing which described the work program planned for the prospective National Biodefense Analysis and Countermeasures Center (NBACC), particularly for one of its four sub-centers, the Biothreat Characterization Center (BTCC), became available. It was proposed that studies be carried out in 16 different subject areas, of which the following nine seemed particularly significant: genetic engineering; susceptibility to current therapeutics; host-range studies; environmental stability; aerosol animal-model development; aerosol dynamics; novel packaging; novel delivery of threat; bioregulators and immunomodulators; and "*Red Teaming*," that is, duplication of threat scenarios.

In addition, task areas for *biothreat-agent* (BTA) analysis and technical-threat assessment were summarized as "Acquire, Grow, Modify, Store, Stabilize, Package, and Disperse." Classical, emerging, and genetically engineered pathogens were to be characterized for their BTA potential. Aerobiology, aerosol physics, and environmental stability would be studied in wet-laboratory and computer-laboratory settings. "Computational modeling of feasibility, methods, and scale of production" would be undertaken, and "*Red Team*" operational scenarios and capabilities would be assessed. BTA use and countermeasure effectiveness would be studied "across the spectrum of potential attack scenarios" through "high-fidelity modeling and simulation." Biological Weapons Convention states: Each State Party to this Convention undertakes never in any circumstances to develop, produce, stockpile or otherwise, or retain:

1. *Microbial* or other *biological agents* or toxins, whatever their origin or method of production, of types and in quantities that have no justification for prophylactic, protective, or other peaceful purpose.

2. *Weapons*, equipment, or means of delivery designed to use such agents or toxins for hostile purposes or in armed conflict.

In order to assure that the activities of all U.S. Government entities remained within the bounds of Article 1, on the ratification of the BWC in 1975 the White House issued the so-called "*Scowcroft Memorandum*" on December 23, 1975.

As an indication of both the ambiguity and confusion surrounding the question of "*offensive*" and "*defensive*" biological weapons relevant research, the following official U.S. policy statements are important to note. A very brief U.S. DoD press statement on January 8, 2002, on *Nuclear*, *Biological*, and *Chemical Warfare* Defense answers the question, "Is the U.S. still developing biological weapons to use against our enemies?" The answer provided began: "As required by executive order, the U.S. Government ceased all offensive biological research in November 1969 ..." However, the original 1969 U.S. policy decision is worded rather differently. The operative paragraph of National Security Decision Memorandum 35 of November 25, 1969, reads:

The United States bacteriological/biological programs will be confined to research and developments for defensive purposes (immunization, safety measures, etc.). This does not preclude research into the offensive aspects of *bacteriological/biological agents* necessary to determine what defensive measures are required.

The analytic study that supported the U.S. policy decision also included a very important relevant paragraph. In response to the question, "Should the U.S. maintain only a RDT&E program," it replied: There are really two sub-issues here: (1) should the U.S. restrict its program to RDT&E for defensive purposes only, or (2) should the U.S. conduct both offensive and defensive RDT&E? While it is agreed that even RDT&E for defensive purposes only would require some offensive R&D, it is also agreed that there is a distinction between the two issues. A defensive purposes only R&D program would emphasize basic and exploratory research on all aspects of BW, warning devices, medical treatment, and prophylaxis. RDT&E for offensive purposes would emphasize work on mass production and weaponization and would include standardization of new weapons and agents.

At least through 1989, DoD considered studies which produced more virulent pathogens, sought to stabilize them, or studied dissemination methods, to be characteristic indicators of *offensive* BW research, and explicitly prohibited by the BWC. This was stated by Colonel David Huxsoll, then director of USAMRIID, in testimony to the U.S. Senate Committee on Governmental Affairs in May 1989, and clearly displayed in a diagrammatic schema attached to his testimony.

In 1989, after court action requiring a programmatic review of its Biological Defense Research Program, the United States recorded a decision to continue the Program, stating that it "is in full compliance with the *Biological Weapons convention*" and "does not include the development of any weapons, nor does it attempt to develop new pathogenic organisms for any use."

After learning of the contents of the NBACC briefing described above, three veteran observers of biological weapons issues authored a memorandum entitled "*Biodefense Crosses the Line*." The three authors had no knowledge whatsoever of the 1999/Lederberg/CIA-NAG experience. The memorandum argued four points:

1. That taken together, many of the activities itemized within the NBACC/BTCC research program—most particularly the "Store, Stabilize, Package, and Disperse" sequence and the "Computational modeling of feasibility, methods, and scale of production"—may constitute *development* in the guise of threat assessment. Development is prohibited by the Biological Weapons Convention.
2. That they very likely would be interpreted that way by at least some other states.
3. That U.S. intelligence agencies would judge a BW research program of this character and magnitude found in any other state to be an offensive BW program.
4. That the program would stimulate analogous efforts in other states; in other words, the BW arms race that Tucker and Dando warned against.

Individuals within the current administration have agreed with the third point in conversations among themselves. The response to this critique of the NBACC/BTCC research program by an official of the DHS in 2005 was a barely qualified self-incriminatory admission of the charge: *Homeland Security* officials defend the *biodefense program*, saying that it will be as open as possible and will not breach the *biowarfare agreement*. But they concede that biodefense today, by necessity, requires stretching research boundaries beyond what would have been acceptable before the *anthrax attacks*. "If you have a bad guy who is trying to hurt you with a bioweapon, you have to understand how much material it will take to do harm, what kinds of packages he'll use to keep it stable, how he might deliver it, and how effective it will be," says Maureen McCarthy, director of the Office of Research and Development in the Homeland Security Department's science and technology division. "Those are hard questions. You can't answer them in a vacuum." There is no "*vacuum*;" of the BWC fills that space. And one cannot "stretch" international treaties to which one is a state party. A recent report on NBACC prepared for Congress by the Congressional Research Service unfortunately includes no discussion of these issues.

On April 28, 2004, at the conclusion of a year's review, the Bush administration disclosed details of the new National Biodefense Directive. Among them, reportedly, was that "the U.S. intelligence community is under orders to carry out studies examining the types of genetically engineered '*bugs*' terrorists could be working on to mount an attack." The intelligence community is not the place that such research should be carried out, if it should be carried out at all. Biodefense is the mission, all or in part, of a sufficient number of other U.S. Government agencies and facilities, which are perfectly capable of carrying out whatever tasks are necessary. These include:

- USAMRIID (DoD)
- Dugway (DoD)
- The Centers for Disease Control and Prevention (CDC)
- Walter Reed Army Institute of Research (DoD)
- Naval Medical Research Institute (DoD)
- DARPA (DoD)
- Edgewood Chemical Biological Center (ECBC) (DoD)
- DTRA (DoD)
- Department of Energy (DOE) laboratories
- Department of Agriculture
- Environmental Protection Agency and now, even the National Institutes of Health (NIH)

The CIA can obtain any information regarding biological agents that it needs in order to carry out its legitimate activities in the sphere of U.S. national security from these other U.S. agencies or organizations. Placing molecular genetics research under the jurisdiction of the intelligence community guarantees that there will be no independent oversight of it.

In the last year or two, there have been a series of major national and international reports that have identified lines of "dual use" molecular genetics research which are particularly problematical. In essence that means that they could be misused to develop more advanced biological weapons. Each of these reports suggests that such lines of research should be subject to particular and special formal oversight by one or another mechanism, much more oversight than presently exists anywhere. Although none of these explicitly engage the question of BWC treaty compliance, it is interesting to match the nine NBACC program elements quoted earlier with the research groupings in these reports. Two German researchers, Nixdorff and Bender, were the first to produce a short compilation. In discussing "modifications of microorganisms that might have significance for bioweapons, [they] identified four classes of microbial manipulations that have been the subject of intense debate within and outside the scientific community: (i) The transfer of antibiotic resistance to microorganisms; (ii) Modification of the antigenic properties of microorganisms; (iii) Modification of the stability of microorganisms to the environment; and (iv) The transfer of pathogenic properties to microorganisms."

A few months later, DoD's Defense Threat Reduction Agency convened a workshop to consider possible limitations on the publication of scientific research that "could be misused for *biological warfare* or *terrorism*." It considered research that aims to achieve one or more of six weaponization-related goals.

1. Enhance pathogen infectivity, pathogenicity, *antibiotic resistance*, or resistance of host immunological defenses.
2. Improve the ability of a microbial pathogen to remain viable and virulent during prolonged storage and/or after release into the environment.
3. Facilitate the dissemination of biological agents as a fine-particle *aerosol*.
4. Facilitate the dissemination of a *biological agent* by contamination of food or water sources.
5. Create a novel pathogen or one with characteristics that have been altered to evade current detection methods or host *immune defenses*.
6. Assemble oligonucleotides to synthesize the genome of a *pathogenic microorganism*.

The third compilation appeared in a 2004 report prepared by a Committee of the U.S. National Academy of Sciences. The Committee identified seven classes of experiments that it believes illustrate the types of endeavors or discoveries that will require review and discussion by informed members of the scientific and medical community before they are undertaken or, if carried out, before they are published in full detail. They include experiments that:

1. Would demonstrate how to render a *vaccine ineffective*.
2. Would confer resistance to therapeutically useful antibiotics or *antiviral agents*.
3. Would enhance the *virulence* of a pathogen or render a nonpathogen virulent.
4. Would increase transmissibility of a pathogen.
5. Would alter the host range of a *pathogen*.
6. Would enable the evasion of diagnostic/detection *modalities*.
7. Would enable the weaponization of a biological agent or *toxin*.

In December 2004, a report of WHO included a relatively similar group of six:

1. Facilitate the production of toxins that were previously difficult to acquire on a large scale;
2. Make a pathogen resistant to the immune system or to *antibiotic treatment*, hence rendering defensive measures in effective;
3. Modify the environmental stability of a pathogen;
4. Create bacteria and viruses of greater virulence or render previously harmless organisms pathogenic;
5. Change host specificity of microorganisms; or,
6. Render the identification and detection of *engineered pathogens* very difficult (e.g., to bypass the current detection technologies).

A fifth group of experimental categories were elaborated under a University of Maryland project aimed at designing a national and international mechanism for oversight of such work. It divides projects that would be considered "*dangerous*" into three groupings: extremely, moderately and potentially dangerous. The overlap between the nine NBACC program elements quoted earlier and all of these five lists is obvious.

On April 18, 2005, the DHS proposed to categorize the NBACC facility as a Federally Funded Research and Development Center (FFRDC). Historically, a very large portion of the work carried out at FFRDCs, such as the Lincoln Laboratories, the Mitre Corporation, the Center for Naval Analysis, the Institute for Defense Analysis, the three U.S. Department of Energy laboratories, the RAND Corporation (Air Force), and others is classified. As of 2005, there are 37 FFRDCs altogether. In the words of a May 2005 Congressional Research Service report, "Many FFRDCs conduct research principally in classified fields for the *Defense* and *Intelligence Communities*." Whether the transformation of NBACC into a FFRDC will make oversight of its work significantly more difficult than it would otherwise have been in any case, no one can say, but it probably can be assumed that it will. It is almost certain that all of the "*threat assessment*" research that it will undertake will be classified. NBACC's initial building cost is currently estimated at $128 million. In addition to NBACC, the DHS has now announced plans to build a National Bio and Agrodefense Facility (NBAF) at an initial projected building cost of $451 million. It would feature large animal BL-3 and BL-4 research capability in order to work with foreign animal diseases. The insertion of the NBAF into the DHS budget proposal was made by senior administration officials. In May 2005, the Army proposed a $1 billion expansion to USAMRIID to be built at the National Interagency Biodefense Campus at Fort Detrick, Maryland. With inevitable cost overruns by the time these three new facilities will all be completed, the infrastructure investment for them will very likely exceed $2 billion. In addition, the U.S. Department of Agriculture is modernizing the National Animal Disease Center in Ames, Iowa, at a cost of an additional $406 million.

A new element, at least as problematic if not more so than the NBACC research program alone but certainly associated with it, is the newly announced wording by the DHS on the programmatic mission of NBACC. It clearly moves U.S. Government policy in the same—wrong—direction. The NBACC research program does so by implication; this does so in a more formal way. The DHS has apparently adopted and begun to publicly use its own novel interpretation of the provisions of the Biological Weapons Convention. In the FY 2006 Congressional Justification for DHS, the DHS wrote the following in discussing NBACC: The work in these laboratories will be for defensive purposes only. The Convention on the Prohibition of the Development, Production, and Stockpiling of Bacteriological (Biological) and Toxin Weapons and on Their Destruction, also known as the *Biological* and *Toxin Weapons Convention* (BWC), prohibits the development, production, stockpiling, and acquisition of offensive biological weapons. The U.S. is a signatory to this treaty, and all activities performed at the NBACC Facility will comply with this treaty and with all other applicable laws.

The insertion of the single word "*offensive*" in front of "*biological weapons*" is a direct contradiction of all the rest of the statement and of all existing international legal interpretations of the BWC. It implies that the BWC does not prohibit the development, production, and stockpiling of "*defensive*" biological weapons. But, the Biological Weapons Convention does not distinguish between "*offensive*" biological weapons, and any other kind. There is no such thing as "*defensive*" biological weapons. Whatever military doctrine may say regarding distinctions between offensive and defensive conventional weapons, this does not apply to biological weapons. Biological Weapons Convention allows the growth of laboratory quantities of pathogens (agents) for defensive purposes, that is, in order to develop vaccines and pharmaceuticals, test rapid detection systems, masks, decontamination systems and so on. However even the "*development*" of the pathogen is explicitly forbidden – "never in any circumstances" – as is production and stockpiling. Blanket prohibition on the development and production of "weapons, equipment, or means of delivery designed to use such agents or toxins for hostile purposes or in armed conflict." The three qualifying words—*prophylactic*, *protective*, and *peaceful* – which operationally define "*defensive*" in the treaty do not apply to "weapons, equipment, or means of delivery".

DHS's choice of language in its FY2006 budget request was not an accident, but a deliberate, considered decision. The new DHS BWC treaty interpretation first appeared in the September 17, 2004, DHS submission for comments on the Draft Environmental Impact Statement (EIS) for NBACC, and was repeated in the final EIS released by DHS on December 23, 2004. In between the two submissions the Center for Arms Control and Non-Proliferation in Washington, DC had questioned the DHS treaty interpretation, pointing out that in the 1989 Programmatic Environmental Impact Statement for the DoD Chemical and Biological Defense Program (CBDP), DoD states that the Biological Weapons Convention "makes the clear distinction between defensive and offensive efforts by identifying the development of biological weapons delivery systems as a discrete and prohibited activity." In addition, U.S. statutory law, the 1989 implementing legislation for the Biological Weapons Convention makes no distinction between "*offensive*" and "*defensive*" biological weapons. DHS responded to these points by stating in its final EIS that "the DoD's programmatic NEPA documentation cited has been superseded." (NEPA refers to the National Environmental Policy Act, under which federal agencies are required to provide Environmental Impact Statements.) In essence, the formal DHS response in this instance was the same as the informal comment by DHS's Maureen McArthy quoted earlier:

that the times had changed. The DHS phrase about "superseded" brashly presumes to revoke the DoD legal commitments made in 1989, and the interpretation of Article 1 of the BWC that the U.S. Government presumably held between 1972 and certainly at least until 1989, and as best as anyone knows until 2004. The times may very well have changed, but the provisions of the Biological Weapons Convention have not, and under international law, they are not open to being changed by unilateral interpretations.

An analogous situation occurred some 20 years ago involving another arms control treaty. In the mid-1980s the Reagan administration attempted to reinterpret one of the provisions of the ABM Treaty through a memorandum—the so-called "*Sofaer amendment*"—written by Abraham Sofaer, Legal Advisor to the Department of State. It blandly stated that the United States could carry out a particular category of ABM testing which the ABM Treaty forbade. In that instance, the U.S. Senate took objection, stating that since it was the duty of the Senate to ratify international treaties, the Office of the President had no authority to modify the terms of the treaty once it had been ratified. Extensive Senate Hearings were held, and books and monographs were written on the subject. The USSR, the treaty cosignatory with the United States, also stated that it would not accept the suggested modification or the activities, and that the integrity of other U.S.-USSR strategic arms limitation agreements would be placed in jeopardy if the United States unilaterally began the testing that it proposed to do. The Reagan administration ultimately withdrew its proposed unilateral treaty modification.

These issues have certainly come to the attention of working level officials in the Foreign Ministries of at least some countries closely allied with the United States, but it is not known whether it has become a matter of private diplomatic discussion between those countries and the United States. Probably not. The subject has not been broached publicly. The entire area of oversight of problematical "*dual use*" research in molecular genetics and its applications in the United States appears to range from inadequate at local levels to virtually nonexistent at the national level and in terms of BWC treaty compliance.

In the late 1970s, the NIH established a system applicable to any institution within the United States or overseas that receives any support from NIH for recombinant DNA research. That system required every such institution to establish an Institutional Biosafety Committee (IBC), and to register that IBC with the NIH Office of Biotechnology Activities. The IBCs were then to oversee projects within their own institutions following a common set of guidelines and criteria. Failure to adhere to the NIH guidelines could result in the suspension, limitation, or termination of NIH funding for recombinant research. However, this system does not apply except on a voluntary basis to a very large population of institutions that are not recipients of NIH funding, such U.S. Government biodefense laboratories and U.S. Government contractors, as well as hundreds of commercial biotechnology enterprises. For example, the Battelle Memorial Institute, a government contractor which has been carrying out Project Jefferson at least since 2001, the development of a vaccine resistant strain of anthrax to duplicate Soviet-era work, apparently has no functioning IBC, although the research is by definition recombinant DNA work. The Departments of Energy, Defense, and Agriculture have numerous facilities carrying out recombinant DNA research at the same time as they have no NIH-registered IBCs. No one knows how much *biodefense* and *bioterrorism* research falls entirely outside the NIH guidelines and the IBC system. In addition, a recent study demonstrated that a very large proportion of the IBCs operated in haphazard fashion or not at all. Some three dozen commercial entities that do receive NIH funding for biodefense research had no IBC registered with the NIH at all. Finally, the existing NIH guidelines do not currently include consideration of the security or proliferation implications of dual-use research, and it is for that reason that the National Academy of Sciences committee recommended that the seven categories of experiments of concern be added to the NIH oversight process.

In response to the recommendations of the NAS committee report, the administration announced the establishment of a National Science Advisory Board for Biosecurity (NSABB) on March 4, 2004. Its mandate was to last for 2 years. The NSABB was established by the Department of Health and Human Services (DHHS) and is to be housed within the National Institutes of Health (NIH). The staff of the NSABB was not appointed until 11 months later in February 2005. No announcement of the membership of the Board was made until its first meeting on June 30, 2005. The responsibilities of the NSABB obviously were not seen as a significant priority by the administration. A May 30, 2005, announcement stated that "The Board is charged with advising on the development of: guidelines for the oversight of dual-use research; national policies governing the publication and communication of sensitive research results ..." The apparent notion was that the NSABB would oversee a process by which the National Academy committee's suggestions regarding experiments of "*dual use*" concern would be grafted on to the IBC system. The NSABB will not itself review individual research project protocols; it will only respond to requests for guidance. That the local institutional committees would perform this much more substantive task, given their highly problematic record mentioned above in dealing with much less demanding considerations, seems dubious at best. A very similar opinion was expressed by the chairmen of six IBCs at major universities in the southeastern United States. But most importantly, the NSABB is to have no oversight over classified BW-relevant research, which is the location in which the most problematical dual-use research is likely to take place.

As an example of the contradictory and counterproductive nature of even current NIH funding decisions, the NIAID at NIH recently announced a grant for a researcher hoping to develop a method of countering the action of botulinum toxin. The grant recipient, Dr. Kim Janda, noted that the task would be difficult because "the *neurotoxin* [is] quite unstable." He would therefore be "collaborating with scientists in Wisconsin to develop a more stable form of the neurotoxin, one that is more easily studied." The biological weapon programs of both the United States and the USSR discovered, independently, that *botulinum toxin* was not a very useful *biological weapon* because the toxin becomes unstable with increased purification. Here was an NIH research award whose purpose was announced as "aimed at stopping *bioterrorism weapons*" that was proposing to develop "a more stable form of the toxin" in order to carry out its research. However, that means preparing a much more effective botulinum toxin than had been available before, which had evaded preparation by the two largest offensive biological weapons programs in history, except in very small quantities.

Another example is the proposal that, better developed animal models and studies of the aerobiological properties of filoviruses [*Ebola* and *Marburg virus*] need to be conducted—which are critical to evaluating the threat posed by filoviruses.... Without data, there can be little understanding of the level of threat that filoviruses present. For example, it is not clear from the available data whether filoviruses would cause large-scale infections and deaths if disseminated by aerosol over a city without extensive preparation or modification (*weaponization*).

This is a perfect example of research designed to probe a potential vulnerability, threat assessment in the absence of a verified threat. As the authors clearly state in their paper, filovirus infection in naturc does not occur via aerosol. The research therefore breaks new ground, and innumerable examples could be proposed all of which would, in effect, be pushing into areas that may not be justifiable.

It is clear that at the time of the three classified biodefense projects in the 1999-2001 time period, there was no U.S. Government-wide NSC-interagency process to review the compatibility of all elements of the U.S. biodefense program with Article I of the BWC. Six years later, there still is none. Even existing mandated treaty compliance frameworks, such as within the U.S. DoD, to carry out reviews of BW R&D projects did not function in the case of its own classified project in 1999-2001. DoD Directive Number 2060.1 mandated the establishment of DoD Compliance Review Groups (CRGs). An analogous directive existed in years prior to 2001. No CRG review of the DoD project (BACUS) ever took place. Whether one took place within the DIA, whose project it was, is unknown. Nor was information about the project ever brought to the attention of the *National Security Council* (NSC).

There is no functioning overall U.S. Government compliance oversight process today for research that impacts the BWC. DoD continues to oppose the formation of such an NSC-level treaty compliance process. Others have suggested that a U.S. BWC compliance review process should be located in the National Academy of Sciences. That may additionally be desirable as a backstop, but it is a government function, and it belongs at the NSC level. With a second heavyweight Cabinet department, Homeland Security, reinterpreting the most critical provisions of the BWC, presumably to permit research that was heretofore considered out of bounds, the picture becomes problematical and dangerous.

The cost of a U.S. BW research program that may cross into "*development*," and of potential U.S. noncompliance with the BWC, very simply means the weakening of the BWC and the international regime that stands in the way of the proliferation of BW to new states or to nonstate actors/terrorist groups. That is certainly not something that anyone who wants such proliferation halted and reversed can possibly want to happen. On May 2, 2005, the U.S. Department of State released a Fact Sheet on "United States Initiatives to Prevent Proliferation." It described seven unquestionably desirable U.S. Government programs. It did not, however, so much as mention any of the international nonproliferation treaty regimes: the Nuclear Non-

Proliferation Treaty, the Chemical Weapons Convention, or the Biological Weapon Convention. This emphasis was repeated at the end of May in statements by President George W. Bush and Secretary of State Condolezza Rice on the occasion of the second anniversary of the U.S. Proliferation Security Initiative. The current U.S. administration does not show much sympathy for international regimes.

It is sometimes said that criticism of the United States and its policies in regard to the BWC itself weakens the Convention, and serves to give other nations that may have no interest at all in observing its provisions a cover for their own misbehaviour. There appears to be no way to address major issues without introducing such risk. Caution and reconsideration only rarely take place within the government, and if the trend appears to be a deteriorating one, the issues have to be raised publicly. Unfortunately, the United States also loses a portion of its leverage to raise questions about possible questionable activities in other states.

INDEX